PRAISE FOR OTHER BOOKS BY GUY P. HARRISON

50 POPULAR BELIEFS THAT PEOPLE THINK ARE TRUE

"What would it take to create a world in which fantasy is not confused for fact and public policy is based on objective reality? I don't know for sure. But a good place to start would be for everyone on Earth to read this book."
—Neil deGrasse Tyson, astrophysicist,
American Museum of Natural History

"Harrison has added to the growing body of skeptical literature a contribution that will continue to move our culture toward one that openly embraces reason, science, and logic."
—Michael Shermer, publisher of *Skeptic*,
columnist for *Scientific American*,
and author of *The Believing Brain*

"Being a skeptic can be hard work, but Harrison makes it a lot easier. . . . This is the book I wish I had written."
—Phil Plait, author of *The Bad Astronomer*

"Deserves to be shelved alongside the works of such giants of the field as Randi, Shermer, Kurtz, and Nickell. . . . A valuable, not to mention very entertainingly written, addition to the literature of skepticism."
—*Booklist*

50 REASONS PEOPLE GIVE FOR BELIEVING IN A GOD

"Deep wisdom and patient explanations fill this excellent book."
—James A. Haught, editor of the *Charleston Gazette*

"Engaging and enlightening. . . . Read this book to explore the many and diverse reasons for belief."
—Michael Shermer, publisher of *Skeptic*,
columnist for *Scientific American*,
and author of *The Believing Brain*

"Doesn't bully or condescend. Reading Harrison's book is like having an amiable chat with a wise old friend."

> —Cameron M. Smith and Charles Sullivan,
> authors of *The Top 10 Myths about Evolution*

50 SIMPLE QUESTIONS FOR EVERY CHRISTIAN

"Guy P. Harrison's new book is a sober, thoughtful, and engaging series of inquiries for us Christians. Answering them or at least responding to whether they are 'simple' or the 'correct' questions is the kind of challenge we should embrace wholeheartedly."

> —Rev. Barry Lynn, author of *Piety & Politics*
> and *The Right to Religious Liberty*

"A thoughtful, conversational, and eminently engaging book. Harrison offers a respectful and yet frank and undauntedly critical approach to Christianity. Believers and nonbelievers alike will find this a worthy, provocative read."

> —Phil Zuckerman, PhD, professor of sociology
> and secular studies, Pitzer College,
> author of *Faith No More*

"Carl Sagan taught us that we make life significant by the courage of our questions and the depth of our answers. Guy P. Harrison offers fifty such questions for Christians to better know their own religion."

> —Cameron M. Smith, PhD,
> Portland State University anthropologist,
> author of *The Fact of Evolution*, and
> coauthor of *The Top 10 Myths about Evolution*

"Award-winning writer Guy P. Harrison is an indispensable voice for science and reason. I sincerely hope every Christian reads this remarkable and important book so that they will better understand the real reasons why so many people are skeptical of Christianity. As Harrison skillfully shows, it is not about anger, arrogance, or rebellion, it's about looking for answers to simple questions."

> —Peter Boghossian, instructor of philosophy,
> Portland State University

"Harrison has posed serious challenges for Christians who explain their religious practice as 'a simple matter of faith.' With persistent, but gentle, words, Harrison injects logic, science, and rational thinking into the discussion of Christian religions, asking only for consideration of facts without the emotional reaction of considering all questions as attacks. This book is another well-written and well-thought-out work by Guy P. Harrison, and it deserves serious consideration by believers and nonbelievers alike."

—Nick Wynne, author of *Florida in the Great Depression: Desperation and Defiance*

"Guy P. Harrison has done the impossible, he has critically analyzed myth, dogma, and belief in the most respectful and accessible way. Many have tried, Guy has succeeded."

—Jake Farr-Wharton, author of *Letters to Christian Leaders*

RACE AND REALITY: WHAT EVERYONE SHOULD KNOW ABOUT OUR BIOLOGICAL DIVERSITY

"This is a very important, profound, enjoyable, and enlightening book. It should go a long way in helping disprove man's most dangerous myth."

—Robert W. Sussman, professor of anthropology, Washington University; editor of *Yearbook of Physical Anthropology*; and editor emeritus of *American Anthropologist*

"A tour de force that conveys the current science on racial classification in a rigorous yet readable way. A book so clearly written, so elegantly crafted, so packed with nuggets that even those who think they know it all about race and racial classification will come away changed."

—David B. Grusky, professor of sociology, Stanford University

"Guy P. Harrison's well-written and passionate plea for eliminating the idea and ideology of race should be widely read. He has shown that the idea of race not only is contradicted by science but is a social anachronism that should not be tolerated by society in the twenty-first century."

—Audrey Smedley, professor emerita, anthropology and African American studies, Virginia Commonwealth University

"Harrison shows we have a lot to learn and he is a great teacher. Drawing on a wide variety of evidence—the hard data from fossils and DNA, interviews with the victims of racism, and personal experiences—Harrison dismantles the 'race' concept, bolt by bolt. Exposing race as a social illusion and political tool—rather than a biological reality—Harrison forces the reader to consider how they think about 'other folk.' Anthropologists have no use for the race concept, and neither should educated citizens."

—Cameron M. Smith, PhD, Department of Anthropology,
Portland State University

"Guy P. Harrison's comprehensive and engaging book should be required reading for anyone who has thought about the benighted issue of 'race.' It will clear the cobwebs from your head."

—Steve Olson, author of *Mapping Human History:*
Discovering the Past through Our Genes

"Harrison challenges us to scrutinize our views about the reality of race and its social consequences, marshaling impressive data and cogent arguments to support his case against the validity of biological race categories. This is a true work of enlightenment, one man's grass-roots effort to raise our collective consciousness to the absurdity of belief in the notion of race, and to raise awareness of the fundamental unity of humankind."

—George Williamson, PhD, Department of Philosophy,
University of Saskatchewan

"For decades, social and biological scientists have amassed evidence demonstrating that the human species has no races, and that differences between groups called races are not biologically based. Harrison makes this knowledge accessible, and knocks the props out from under scientific arguments that have been used to justify racism."

—Jefferson M. Fish, professor emeritus of psychology, St.
John's University, New York

Visit the author's website at www.guypharrison.com

Think

Guy P. Harrison

Think

Why You Should Question Everything

 **Prometheus Books**

59 John Glenn Drive
Amherst, New York 14228–2119

Published 2013 by Prometheus Books

Prometheus Books recognizes the following registered trademarks, trademarks, and service marks mentioned within the text: Epcot^{SM}, Google®, Scotch®, Wikipedia®, and YouTube®.

Cover image © 2013 Media Bakery
Cover design by Nicole Sommer-Lecht
Illustrations by Kevin Hand, © Guy P. Harrison

Inquiries should be addressed to
Prometheus Books
59 John Glenn Drive
Amherst, New York 14228–2119
VOICE: 716–691–0133 • FAX: 716–691–0137
WWW.PROMETHEUSBOOKS.COM

17 16 15 14 13   5 4 3 2 1

Library of Congress Cataloging-in-Publication Data

Harrison, Guy P., author.
    Think : why you should question everything / Guy P. Harrison.
        pages cm
    Includes bibliographical references and index.
    ISBN 978-1-61614-807-2 (pbk.)
    ISBN 978-1-61614-808-9 (ebook)
    1. Reasoning—Miscellanea. 2. Critical thinking—Miscellanea.
3. Skepticism—Miscellanea. 4. Science—Methodology. I. Title.

BC177.H378 2013
153.4—dc23
                                                                    2013024867

Printed in the United States of America

This book is dedicated to my mother, Coni Harrison.
Thanks for all those trips to the public library, Mom. They mattered.

CONTENTS

2. PAY A VISIT TO THE STRANGE THING THAT LIVES INSIDE YOUR HEAD

ACKNOWLEDGMENTS

Thank you to my wife and children for encouraging me and filling my life with love. I am also grateful to the brilliant and supportive team at Prometheus Books, especially Steven L. Mitchell, Jade Zora Scibilia, Meghan Quinn, Jill Maxick, Catherine Roberts-Abel, Nicole Sommer-Lecht, Bruce Carle, and Brian McMahon. And thank you to Kevin Hand for the artwork that appears at the start of each chapter in this book. He is one of the world's best illustrators and working with him is always a joy.

INTRODUCTION

My life has been better because of a key decision I made many years ago. I'm not sure exactly when it happened or if it even occurred in a single moment, but somewhere along the way I decided to be a good skeptic and think like a scientist every day. This simple but crucial choice has saved me time, money, and probably a lot of embarrassment and regret, too. Who knows? It might have saved my life. I didn't take this path because I wanted to feel superior or look down on others for believing things that probably aren't true. Skepticism is not arrogance. I didn't make the decision because I wanted a cold and unemotional life devoid of mystery. Skepticism does not drain the joy from life; in fact, it can add much to it. Thinking like a scientist doesn't mean one can't enjoy fiction and fantasy and dream impossible dreams. Thinking skeptically certainly does not require one to deny or ignore the many unanswered questions all around us. I simply thought that it made sense to embrace skepticism because I wanted to minimize the amount of my life I wasted on lies, mistakes, and misperceptions. I also didn't want to run the risk of being a sucker or easy prey for con artists.

If I hadn't been skeptical about every unusual and important claim that came along over the years, how many crackpot medical treatments might I have squandered my money on? Would one or some of them have harmed my health? It's impossible to know, but I'm glad I didn't risk it. How many smooth-talking crooks and seductive but worthless products failed to fool me? How many organizations based on lies or hopelessly dubious claims did I not join, thereby not wasting precious hours, days, and years of my life? How much stronger and more confident have I been when facing each new day because virtually all my fears and anxieties stop where known reality ends?

I worry about those who are weak skeptics. I feel sad when I see people stumble. There shouldn't be so many victims. I believe anyone can think like a scientist and everyone should want to. This book will show you how. If you are already a good skeptic, it can help make you a better one. It will also show you more effective ways in which you might spread the word among friends and family members who may be a little too loose with their beliefs. Thinking like a scientist doesn't guarantee anyone a perfect life that is without mistakes and free from danger, of course. But it does put you in the best possible position to successfully navigate your way through this crazy world of ours. Insufficient skepticism is perhaps the most unrecognized and underreported global crisis of all. Applied vigorously and consistently, skepticism could change humankind for the better overnight. It is our most neglected defense and underutilized weapon. Politicians never mention it. Teachers rarely teach it. Few parents encourage it. So, with little resistance, the world keeps spinning its lies and delusions, generation after generation. The cost of leaving so many unproven claims and bogus beliefs unchallenged can be seen in diminished and lost lives. Most people underestimate how easy it is for healthy and bright people, including themselves, to be seduced by the irrational ideas of others or led to fantasyland by the natural processes of their own brains. One of the goals of this book is to make sure you don't make that mistake.

This may be described as an instructional or educational book, I suppose, but it is very important for readers to recognize that it is not a collection of facts or a presentation of arguments meant to instruct people on *what* to think. This book is about *how* to think. I'm not looking for loyal teammates or hoping to enlarge the choir that I preach to. I want to inspire readers to think for themselves in sensible ways that allow them to chart their own course in life without being tripped up by one irrational belief after another. I absolutely do not want readers to look to me as some voice of authority that they can rely on as a shortcut to their own thinking. For example, I don't think psychics and mediums can read minds and carry on two-way conversations with dead people. But don't take my word for it. Figure it out for yourself like a good skeptic would. I've been wrong before; maybe I'm wrong about this.

This book is a call for *self-reliance*. It's about *you*. In order to live in the real world and hopefully avoid becoming the pathetic plaything of crooks, kooks, and sincere-but-deluded people, you have to use and rely on *your* brain, no one else's. Sure, expert opinion and information from credible sources matter, but not nearly as much as your own ability to react wisely when an extraordinary claim knocks on your door. For your own sake, you need to know what to do so that you can determine whether or not you should let it in or slam the door shut.

Have no doubt, there are an infinite number of weird claims, unusual ideas, dangerous ideas, and unlikely-to-be-true beliefs stalking you every day. People make up new claims every day, so we can't possibly carry around in our heads tailored, preloaded responses for each one. A good skeptic is ready for all of them simply by knowing how to think critically and by understanding the wisdom in questioning everything. A good skeptic also needs to know basic facts about what the human brain needs in order to work well, because it all starts there. Researchers continue to accumulate evidence that show nutrition and physical activity are more important than we ever knew. When spreading the word about science and skepticism to others, it also helps to know a minimal amount of background information and alternate explanations for some of the more popular beliefs. This book will help with all that, too.

Skepticism is an important issue for everyone. It's something we all need, regardless of intelligence, education, location, social status, or income. The mere fact that you have a human brain makes you extremely vulnerable to believing things that are not true. There is no escaping this. The natural processes and instincts we all have set us up to fall for things that are not true or real. It comes with being human. If you deny this and think you are immune, it only makes you more vulnerable to nonsense and delusion.

Fortunately, the same brain that so often fails us and tricks us can also be very good at protecting us. It can be the most vigilant and effective guardian you could ever have. Or it can be a source of constant problems. It's up to you.

—Guy P. Harrison
Earth

Chapter 1

STANDING TALL ON A FANTASY-PRONE PLANET

Generally I like to assume that it's not smart to generalize and make assumptions. Too many things fall somewhere into that broad abyss between black and white. Since you are reading this book, however, I'm going to go out on a limb and make two guesses. One, I'm willing to bet that you are a human being and, two, you live on Earth. If I'm right, too bad for you. Whether or not you have realized it, your address and your membership to the human species have condemned you to a life sentence atop the crust of a spherical madhouse. Through no fault of your own, you have been born into a world teeming with good people who mean well but to one degree or another are deranged, deluded, and just plain wrong. And most of them are eager to lure you into the fog with them. But wait, there's more.

There is also no shortage of bad people who don't mean well. They trade in lies. Their goals may include harming you, exploiting you, taking your money, or all three. Unfortunately, you can't hide from any of these people. Some days it's the dishonest ones who come for you, and other days it's the sincere but mistaken ones. Sooner or later they will find you; they always find you. In fact, many of them are in your life right now, no doubt. They live in your neighborhood. You go to school with them. You work with them. Sometimes they are close friends. Every family has at least a few. You could try to run, but to where? No place is safe because they are everywhere. No city, no society, is free of them. Again and again they will reach out to you. They have a million methods and a trillion techniques. Their ideas can be like microbial parasites, requiring only the smallest crack or opening to squirm in and take over your brain. *Join us. Pay us. Give us your time. Buy this. Trust this cure. Pay us now, please. Think like us. Be one of us. Believe us. And don't forget to pay us.*

When they come for you, the easiest thing to do in the short term may be to hand over your brain or your wallet, and then march quietly into the swamp of muddy thinking like a good little drone. The long-term consequences for you may be harsh, however. Perhaps some people are so weak and so passive that assimilation really is their only option. Fortunately, this is not the case with you, I'm guessing. You would probably like to hold onto your brain, money, health, and dignity. Am I right?

I wish this book were absolutely universal in appeal, but it might not be the best fit for everyone. Therefore, let's address a few basic points early to make sure that this book is right for you:

- If you like the idea of spending a lifetime stumbling back and forth between a variety of unproven claims and strange beliefs that are almost certainly not true, then stop reading this book now.
- If you think it would be really cool to waste thousands of dollars over your lifetime on bogus miracle cures and absurd products hawked on infomercials, then you need to immediately put this book down and step away from it.
- If worrying about things that don't exist and cannot harm you sounds smart, then run as far away from this book as you can. Hurry, it might attack you!
- If being intellectually smothered within a group or organization that discourages thinking and forbids asking meaningful questions sounds comforting and reasonable, then buy as many copies of this book that you can afford and then burn them all. As they are burning, be sure to chant something. The words don't matter, anything will do.

I'm sorry if you now feel that this book is not for you. But at least your path in life is clear: Trust everything anyone tells you. Live by a creed that says any claim that feels good and sounds good must be true. Forget reason and always go with your heart. Who needs facts when your gut speaks to you? Never second-guess anything. Never doubt, never question, and never worry about things like evidence, proof, logic, or that obnoxious thing called science. That's it; you are ready to live a dream life. Before you begin your journey, however, please close this book and slam it against your forehead ten times. This will give you good luck. Trust me.

Still here? Then it can only mean one thing: You are ready to think! You must be the kind of person who likes the idea of standing tall on Planet Crazy with eyes wide open and a revved-up thinking machine, just like a twenty-first-century human should. You want to be good at spotting scams and recognizing worthless products for sale. You want the ability to see through hollow words and dishonest promises virtually every time. A safer and more efficient life sounds good to you. You are ready to put up a fight against all those barbarians at the gate who are determined to invade your skull. Their mission is to use and abuse you with counterfeit claims and fake philosophies, but you are going to be ready for them.

THINK LIKE A SCIENTIST

The most effective antidote for bad thinking is *good thinking*. The best way to make con artists vanish is to *see* them. The best way to silence crazy claims is simply to *listen* to them with a sharp brain and then ask the right questions. It is important to look at wild claims and hear incredible sales pitches with deliberate effort. You can't be passive about this. Crooks and kooks love finding a brain with a wide-open doorway and nobody standing guard. Respond to extraordinary claims in a way that is similar, in spirit at least, to how a scientist would investigate a new, exotic illness found in redwood trees or newly discovered microbial colonies in the bellybuttons of Latvian fishermen: observe, research, hypothesize (think up ideas), ask questions, experiment, and share your ideas and conclusions about it with sensible people. Rinse and repeat.

Thinking like a scientist is not that difficult. Young children can do it. Old people can do it. High-school dropouts can do it. It doesn't require you to memorize the periodic table of elements or to understand quantum physics. Thinking like a scientist in this context only means that you maintain a healthy level of curiosity and doubt. You aren't afraid to ask questions and request evidence, and you don't draw conclusions about things until you have very good reasons to do so. You also must be willing to change your mind if your conclusions turn out to be wrong.

The bad news is that the people with weird, unproven claims out-number you. They may have popular opinion, tradition, and the always-attractive lure of lazy thinking on their side. You may think common sense, the laws of nature, and logic would constrain these people, but they rarely burden themselves with such concerns because theirs is the domain of the make-believe where rules are invented or discarded as desired. They might also cheat by appealing to emotion and tick-ling your soft underbelly. They might tell you that their way is warm, comforting, exciting, and meaningful while the skeptic's world is cold, lonely, boring, and empty. Don't believe such lies.

Deciding to think like a scientist is the hard part; *doing* it is fairly easy. The mechanics are simple and straightforward: Proof comes before belief. Nothing is ever beyond question or revision. But don't be fooled, this is a lifelong war and every day is a battle. Irrational thinkers and their crazy claims never go away. You can never declare victory and let down your guard because there are always more nearby who are looking for another mind to infect. Dismantle one of these brain-snatching beliefs with your skepticism, and ten more are ready and waiting to take its place. Annihilate one hollow claim with reason, and twenty more rise up. The best you can do is hold them off in a siege that never ends. And through it all, you need to stay positive, to hold onto your humanity. You can't allow yourself to become frustrated and bitter and begin to pull away from your fellow humans. That's not good for you and not good for the world. Excluding the liars and con artists, people who believe in bad ideas are victims of bad thinking, that's all. Despising them or giving up on them is not constructive. I've been pro-moting and writing about skepticism for many years, and I have dealt with some of the most idiotic and depraved lunacy imaginable, and yet I can still say without hesitation that I love people. They frustrate me and disappoint me to no end, but I still love them.

HALFWAY HOME

Now for some great news: You are already a skeptic and a critical thinker. You're halfway there. Whether or not you recognize it, you apply skepticism and critical-thinking skills every day. We all do because everyone is a skeptic to some degree. Think about it—nobody believes *everything*. For example, if some guy tries to sell you a candy bar that has already been opened and you don't recognize the logo on the wrapper, you probably wouldn't just do the deal and start chewing without giving it some thought. Even if he promises you that it is one of the best-tasting candy bars in the world by far and he's offering you a super-special, once-in-lifetime price, you aren't going to bite on the deal or the candy bar without first doing some critical thinking, right? Even if he further tempts you by adding that it has special probiotic-homeopathic properties guaranteed to detox and revitalize your spleen, you still would likely hesitate. You would assess the claim that it's a great candy bar. *Then why haven't I ever heard of it?* You would try your best to analyze the condition of it. *It's open. How do I know a used candy bar is safe to eat?* The phrase, "too good to be true," might come to mind. You would wonder about the claims he made. *What in the heck does 'probiotic-homeopathic' mean? And why does my spleen need revitalizing, anyway?* You probably also would consider anything you know about the seller's character and reputation. *What do I know about this guy? Is he honest and reliable? Is he an expert on spleens? Did any of his former customers die of food poisoning?* Regardless of who you are and how desperate for candy you may be at that moment, you are likely to ask a few direct questions about the candy bar's appearance, mysterious brand, low price, and medicinal properties before buying it. Even if you were desperately hungry, you probably would request a small sample of the candy bar for a quick sniff or taste experiment. See? You're a skeptic. You think like a scientist. But it's not enough to reserve your skeptical powers for encounters with suspicious candy bars only.

None of us automatically believe, accept, or buy every story, claim, and product that comes along. We think. We ask questions. We routinely use our brains to protect us from making bad decisions and allowing ourselves to be misled by incorrect information and outright lies. This is skepticism in action. But while virtually everyone may become a world-class skeptic when faced with a stranger who wants to sell some funky candy, what happens when a trusted friend tells you that she saw a ghost, says she had a horoscope hit right on the mark, or asks you to buy memberships in a group that promises personal peace and financial wealth? How would you react when an authority figure, someone you may have looked up to for years, insists that they have just the medicine you need to feel better? It's only $99.99 and it's "natural," so it can't possibly hurt you, he says. Are you a world-class skeptic during moments like these, too? Or does something cause you to dial it down? This is what happens for most people. They wouldn't buy a used bike without inspecting every inch of it and test-riding it. But they might buy into a hundred different outrageous claims without blinking, based on little more than feeling or a friend's word. It happens every day. But why? Why wouldn't you want to be skeptical about everything, not just from candy bars to bikes, but from healthcare products to horoscopes and beyond? Skepticism is too reliable and too valuable to leave on the shelf. Use it everywhere and every day.

WAIT, WHAT IS SKEPTICISM?

Before we get too far down this road, let's make sure we know what skepticism is. Skepticism and science are really the same thing and work pretty much the same way. Skepticism is just about having a healthy dose of doubt and using reason to figure out what is probably real from what is probably not real. It means not believing you know something before you can prove it or at least make a very good case for it. Skepticism is nothing more than thinking and withholding belief until enough evidence has been presented. It also means keeping an open mind and being ready and able to change your mind when new and better evidence demands it. Proven doesn't mean forever. A good skeptic thinks about the source of a claim and how well it ties in with what we already know about nature and people; and, most important, a good skeptic resists the temptation to make up answers to important questions. Saying "I don't know" is not uncomfortable for a good skeptic. It's routine.

I think of skepticism as science in action. It's the scientific process modified and personalized for everyone to use in everyday life. Why should professional scientists be the only people who get to think straight? We civilians have just as much a right to use rational thinking as they do! I imagine my own skepticism to be something like a personal force field that protects me from invaders. It's like a balled-up fist that punches out bad ideas aiming to do me harm. Who wouldn't want that? If your skepticism force field is powered up and calibrated correctly, then the crooks, time wasters, crackpots, and creeps who want you to join their crazy clubs or buy their junk tend to bounce right off. Sometimes it's dramatic and they get torched like a fly nose-diving into one of those electrified bug-whackers. Most of the time, however, you barely even notice their failed attempts because they are so weak and silly. But if your skeptical force field happens to be weak or turned off from time to time, it can be a very different story. The infiltrators will slip right in and proceed to infect you with their nonsense. Don't miss the seriousness of this. The quality, strength, and consistency of your skeptical thinking will likely have a direct impact on your safety, success, and quality of life. Doing your best to stay free and clear from bad ideas and bad people who would take advantage of you should be a top priority. We may not have to worry about giant prehistoric cats eating us anymore, but that doesn't mean there aren't plenty of other predators looking for you.

Your skepticism can also be either purely selfish or strictly humanitarian in nature. It's up to you. But either way, skepticism is the wise choice. If one wants to make the world better, safer, and smarter, skeptical thinking is the way to go. Why should me-first divas waste one second of their days on nonsense beliefs when it would mean less time to be selfish and self-centered? Become a good skeptic for the betterment of our species, or do it for yourself and to hell with the world. Just make sure you do it. If you aren't sure which camp you belong in, don't worry. Most good skeptics probably have a bit of both in them. I know I do.

NO OFF DAYS

Being a consistent skeptic can be the difference between life and death. The moment we let down our guard is when bad things come rushing in. Ceasing to think independently, analyze relentlessly, and ask the relevant questions is a big, fat green light for bogus beliefs to raid your inner sanctum. Don't underestimate the importance of this. It's tempting to dismiss the threat of bad beliefs as nothing more than fringe silliness about silly people, nothing much to do with you. But this is deadly serious and none of us are ever in the clear, thanks to the brains we are blessed/cursed with. For lack of skepticism and ignorance about how the human brain misleads its host, many millions of people suffer needlessly every day. Millions throw away money on garbage ideas and rip-off products. Millions sacrifice irretrievable hours, days, and years of their lives on hollow claims. As we will see later in this book, many people die prematurely as a result of their weak or inconsistent skepticism.

Sadly, most people turn off their skepticism at the very times they need it most. Everyone may be skeptical to some degree, but the unfortunate reality is that most people are just not on their toes when it counts. All they probably need is a minimal amount of knowledge about critical thinking and a little encouragement to live more reason-based lives. This is why I write books like this. It's why I do all I can to promote science and reason. For me it's a moral issue, and a big one at that. I can't look at all the madness and suffering that comes from weak skepticism and do nothing about it. This is a gigantic problem that doesn't get the attention it should. For example, you aren't likely to see many, if any, big media news reports on the global cost of gullibility or how much violence, poverty, and death is tied to the failure of the world's families and schools to produce critical thinkers. Nonetheless, whether or not we acknowledge it, the kind of weak and inconsistent skepticism that is typical of people today is a massive drain on humanity and a constant source of suffering. It hurts countless individuals while simultaneously slowing progress and undermining prosperity for all, yet it is unrecognized and unspoken of by virtually everyone everywhere.

I promise that if you invest the time to read this book and apply what you learn, you will have a huge advantage over the majority of people in the world. You can be a leader rather than a follower. While others are tripping over their own feet in the dark while looking for ghosts and monsters that they will never find, you can focus on getting real things done. You can make genuine discoveries and perhaps find everything you need to build a good life for yourself, right here in the real universe. You can choose to spend your money on things that actually exist, make sense, and work. Unburdened by the weight of too many delusions, you can devote more time and energy to family and friends or helping strangers in need. As a good skeptic, you are in a better position to utilize the benefits of science and scientific thinking to help you make the most of your life. It's free and simple to decide to embrace skepticism, but the payoff is huge.

WHO CARES? SO WHAT? AND WHY BOTHER?

We live in a wonderful age of awesome electronic gadgets, amazing medical science, and computers that get smaller and more powerful each year. There is even a space station flying above our heads at a speed of more than 17,000 miles per hour. It has been continually occupied by humans for more than a decade. It's the tiniest of footholds, but we can honestly say that our species lives on Earth *and* in space. We are mapping the genomes of plants and animals, mapping our brain, exploring Mars with robots, and learning more about the vast microbial universe all around us every day.

I'm sure you already knew all about that cool stuff, of course. But did you also know that thousands of "witches" are banished from their homes and tortured, and some are killed each year? It's true. In the age of science, millions of us still cling to the same irrational fear of spells and hexes that have haunted us since we first made them up. We may have left the Stone Age, but the Stone Age is still in us. Psychology professor Hank Davis has an excellent book about this problem called *Caveman Logic: The Persistence of Primitive Thinking in a Modern World*.[1] He's right; prehistoric thinking remains alive and well in the twenty-first century. If you are skeptical enough to know better than to be afraid of a witch, resist the urge to giggle here because there is nothing funny about this. This fear of magic and invisible demons that swirls around inside the heads of weak skeptics leads some of them to murder. "Child witches" in Africa, for example, are ostracized, beaten, starved, and killed by Christians who say they are only doing what their god demands.[2] Reportedly it is common for mobs to beat and kill "witches" and "sorcerers" in rural India.[3] People should not be murdered for having extraordinary powers that no one has ever been able to reliably demonstrate or to prove exist in the first place. And people should not ever become murderers because they didn't hear that it's always best to think before believing.

Far more deadly and widespread than witch persecution in Africa and Asia, weak skepticism leads to suffering and/or death for unknown numbers of people everywhere who trust in medical quackery over evidence-based medicine. There can be no doubt that many hundreds of millions of people have suffered or died too soon because they weren't skeptical enough when it came to their personal healthcare or that of their children.

It's impossible to know how much money people have wasted on bad beliefs and highly questionable claims in the last year, decade, or century. Whatever it is, I suspect it would be measured in the trillions of dollars. I have always found it odd that so many people in general are worried about money and eager to get as much of it as they can, yet they give it up so easily to nonsense.

THE GREATEST SHOW IN THE UNIVERSE

Before we go any deeper into what being a good skeptic is all about, let's make sure that one important point is perfectly clear. The kind of skepticism I am promoting in this book is positive, constructive, and optimistic. In my opinion, good skeptics gain far more than they lose. The kind of good skeptic that I hope you will consider becoming and encourage others to become doesn't have to sacrifice excitement and accept a less thrilling life. Sure, believing in unproven claims such as the Bermuda Triangle, Bigfoot, UFOs, and psychics can be a fun ride. But those trips are dull dead-ends compared to what reality offers. Astrology can't compete with astronomy. Not even close. One only has to give skepticism and science a chance to find out. Unproven ESP claims pale in comparison with some of the things that cutting-edge brain science is discovering and doing these days. While charlatans claim to bend spoons with their minds, science actually moves robotic limbs with brain impulses.[4] Looking up and hoping for a flying saucer full of aliens to fly by might be fun for a while, but the scientific search for extraterrestrial life gets more exciting with each passing year as technology improves and we discover more planets that might harbor life. Look back at history. What did magic and monsters achieve? Science and reality do something all those supernatural/paranormal promises never do. Science and reality deliver the goods.

Skepticism is negative? Ridiculous. If anyone near you suggests that, move away from them so you don't catch whatever it is they are suffering from. Skepticism is the brain's perfect diet. It trims away the fat and turns you into a lean thinking machine. It dumps unnecessary ballast and clears the way ahead so that you can get to all the good stuff out there waiting to be discovered. Skepticism frees us so we can dedicate ourselves to real things and real people. Anyone who tries to tell you that skeptics are pessimistic party poopers or that the scientific process is dull and takes the thrill out of life has no idea what they are talking about. Help them see the light if you can, just don't let them drag you down.

The scientific process helps us to discover amazing things about the universe and ourselves every day. One can either be a good skeptic and appreciate science or stumble though life with closed eyes, missing the greatest show ever. This has never been more important than it is today because one risks missing out on more than all the people who ever lived before. Think about someone in the year 10,000 BCE who was determined to think magically and refused or had no interest in discovering much about the world and the universe as it really is. That person's attitude wouldn't have cost them much because relatively little knowledge was available anyway. Science as a formalized process didn't even exist then. No microscopes, no telescopes, no books documenting and explaining centuries of scientific discovery. However, a person who ignores or rejects science in our time turns her back on a staggering amount of important and exciting information. So much is known and so much more is sure to be discovered in your lifetime. Don't miss it because you were too distracted by a bunch of myths and fantasies masquerading as reality.

The deeper you dive into it, the more you realize that science is much more freaky, amazing, wonderful, and important than all that supernatural and paranormal stuff could ever be. Why settle for wild claims that are not supported by good evidence when science has already confirmed so many other wild ideas? Millions of people say that paranormal mind powers can move objects. Big deal; people can say anything. Let's wait until someone gets around to proving it before we get excited. In the meantime, why not check out how nature moves *entire continents*? It's called plate tectonics and scientists have plenty of evidence for it. You can follow the efforts of cryptozoologists with bated breath when they go looking for the Loch Ness monster or you can focus your attention on scientists who actually find new life-forms all the time. There are probably 100 million species out there waiting to be discovered and named. After that, we have to figure out what they do, how they do it, and where they fit into the big picture. There is work to be done, large and small. We have only just begun to scratch the surface of the universe and there is still much more to learn about the atom. Who has time for magic, myths, and lies?

Are you into ghosts? Do you think they are real and important enough to occupy some portion of your mental energy? If not, you probably know someone who is a ghost enthusiast. People who believe in floating, semitranslucent dead people certainly should be free to do so. But for their own sakes I wish they would give science a chance. I never tell people *not* to believe in things. For me, the old bait-and-switch technique is both gentle and effective. I don't mind listening to someone's ghost stories for a while. I might even share my own haunted house story (see chapter 3). But somewhere in the conversation, I'll try my best to interest them in science and reality. For example, why not upgrade from the ghost thing and look into the more likely possibility of parallel universes? It's not conclusive, of course, but there really could be other universes that contain identical or similar versions of you. There may be beings who inhabit other universes in other dimensions who are right next to you right now. Maybe one of them just walked through you. Isn't this possibility freaky enough for anyone's taste? One can also stretch their brain a bit by learning the basics about the structure and behavior of atoms. Reality gets very, very strange at the atomic level. For example, thanks to science, we know that all matter—including us—is mostly empty space. We are far from the solid beings we naturally imagine ourselves to be. Hey, maybe *we* are the ghosts! The nucleus (middle part) accounts for almost all of an atom's matter (solid part). But the nucleus is very tiny compared with the size of the entire atom. If one atom were enlarged to the size of an entire football stadium, for example, the nucleus would be about the size of a golf ball or a marble in the center of the field. The outer electrons would be orbiting way out in the cheap seats. So, back to you, if all the empty space was squeezed out of your body, the "new and condensed you" would be smaller than a grain of rice. (Though it would be a *very* dense and heavy grain of rice). This means, of course, that the next time someone calls you an airhead, you shouldn't be offended but should instead compliment them on their knowledge of atomic structure. But wait, we're not finished. Scientists are now figuring out that all this empty space in us and throughout the universe may not be so *empty* after all. Turns out there seems to be *something* in nothing. See, not only does science answer questions better than anything else does, it also generates new questions better than anything else does.

Is it reasonable to conclude at this time that alien spaceships have visited the Earth? No, because, although it could be true, the best anyone has ever produced to back up this claim are unreliable eyewitness accounts, stories of close encounters, and questionable photos/videos of weird things in the sky. But being skeptical of UFO stories does not mean one can't give serious thought to the possibility of intelligent extraterrestrial life existing somewhere out there. The scientific process has led us to understand how adaptive and persistent life can be here on our planet. That knowledge, coupled with the immense number of opportunities for life that exist in a very large universe, gives us reasonable cause for heightened interest, if not outright enthusiasm. The absence of good evidence means that *believing* in aliens down here doesn't make sense, but *thinking* about aliens out there does.

The skeptical outlook is not a negative mind-set, as some who promote unproven claims would have you believe. Being a skeptic doesn't mean never getting excited about extraordinary claims or dreaming of grand possibilities. Hoping and having wild dreams are natural and healthy. A good skeptic only tries to balance it all with logic and evidence. If I'm being interviewed on some radio show and alien abductions come up, my first reaction is not to snap at a host or caller for believing this unproven and unlikely claim. My immediate instinct is not to *take away* an unfounded belief from the brain of a bad skeptic. I'm more interested in *giving* them something better. So I quickly point out a few basic problems associated with the alien home-invasion claim and then go on to spend as much time as they will allow me to teach something about space exploration and the scientific search for extraterrestrial life. If they fall in love with that, maybe they won't be so impressed with hollow stories.

HATE THE BELIEF, LOVE THE BELIEVER

I always try my best to separate the beliefs from the believers. One can sometimes become frustrated with someone who believes a nonsense claim, especially if it's a destructive belief. But I try to keep in mind that it's the claim that I'm upset with and not the person. When someone gets cancer, you don't get mad at the person, you get mad at the disease, right? In my opinion, good skeptics should avoid being too harsh with people who believe silly things. Be less condemning and more understanding. Teach, don't shout. After all, we've all been there in one way or another and at one time or another. When I was a kid, I believed, briefly, in the ridiculous claim that superintelligent aliens visited the Earth long ago and showed our mentally deficient ancestors how to paint on cave walls, build pyramids, brush their teeth, and so on. I'm not proud of it, but I'm not ashamed of it, either. I'm human. Such stumbles come with the territory.

Like good people, good skeptics care about others and want to offer a helping hand when and where possible. Whether the topic is Atlantis, reincarnation, or conspiracy theories, my goal is to enlighten people, to show them how easy it is for all of us to fall for a good story delivered by a confident storyteller. Simply being a human being and thinking the way we do sets us up to believe these things. I care about the world, so I try to make it less goofy and less dangerous by showing others how they can keep their guard up. Sensible people don't want idiotic ideas wasting valuable space in their heads if they can avoid it. The problem is that they need help to become good at recognizing idiotic ideas when they come along dressed up as perfectly reasonable and respectable claims. If you are not already, when you become a good skeptic you can help others spot the nonsense, too. Simply by spreading skepticism, you can help make the world better, and that's a beautiful thing.

THE YOUNG THINKER

When it comes to adopting a skeptical attitude, now is better than soon. Don't delay in boarding this train. Start getting into the habit of thinking more critically about weird claims and beliefs immediately. Starting today, ask questions and demand evidence before you let yourself accept something unusual or important as true. Being young doesn't matter. A life is no less valuable and precious in the early stages than it is in the middle and later stages. If you are young, don't wait to think straight and think for yourself because it's never okay to waste time and energy on nonsense. Old or young, you want to squeeze everything you can from every moment. Learn, create, explore, have fun with friends, love your family and all that. Don't sacrifice your precious days on somebody else's weird and unproven ideas. Think early and think forever. Why wouldn't you want to? Sure, depending on age and maturity, young people have to trust the older authority figures around them and accept that their own lack of experience means they aren't ready to go it alone. But that doesn't mean young people can't think for themselves to a significant degree, beginning right now. If you are a child or a teenager, do not accept that you are a robot to be programmed entirely by adults. Adults are often wrong, and you have in your possession a powerful brain. Use it. And if you're an adult, don't think it's ever too late to change the way you think. The sooner you can start thinking skeptically, the better.

We are all greatly influenced by our family, friends, and culture, of course, but we can still make the important decision to think independently and respect ourselves enough not to blindly accept everything that the rest of the world tries to shovel into our brains. It's understandable that some young people might feel that skeptical thinking is disrespectful in some circumstances, particularly when it comes up against the beliefs or claims of beloved authority figures such as good parents and favorite teachers. That often happens, of course, so it's a reasonable concern. But keep in mind that there is one form of respect that is even more important than the respect granted to authority figures. It's called *self-respect*. And thinking skeptically is the highest form of self-respect because it means you care about the quality of thoughts and beliefs that inhabit your brain. More than anything, this defines you. Your thoughts and your beliefs determine who you are and who you will be.

The fact is, people in your life may be smart, love you, want only what's best for you—and still be completely wrong about some things. If current circumstances make openly questioning certain claims and beliefs too uncomfortable or dangerous for you, then don't do it openly. But do it privately. Being a good skeptic and thinking freely while keeping your mouth shut is infinitely better than not thinking at all. It may feel lonely at times, but it can be done. No one can know your private thoughts if you don't share them—not even a psychic.

TAKE RESPONSIBILITY FOR YOUR OWN BRAIN

Good parents, teachers, and role models are quick to lecture young people about the importance of hard work, education, good manners, not breaking laws, avoiding drug and alcohol abuse, safe driving, being honest, and so on. But how many take the time to explain the importance of skepticism? How many fathers sit with their daughters and teach them how to intellectually burn down the hollow claims of mediums, medical quacks, and astrologers? How many mothers make time to explain to their sons that the normal, everyday workings of the human brain can lead us astray when we try to separate reality from illusion and delusion? Unfortunately, most children grow up and enter adulthood unprepared for what they will face in the form of pseudoscience (false science), unproven claims that aren't worthy of acceptance, and just plain lies. I'm not bashing parents or teachers. Those are difficult and important jobs. I know because I have been a teacher and I am a parent. Besides, in most cases, it's not entirely the parents' fault because they probably never had anyone explain any of this to them when they were growing up, so they don't recognize the need or even know how to teach good thinking skills. Teachers may be hindered from encouraging skeptical thinking because they are often not allowed to deviate from some master plan that demands memorization and regurgitation exclusively from students.

If you are a young person, the bottom line is that thinking like a scientist every day is so important to your chances of having a safe, productive, and fun-filled life that you can't afford to sit back and rely on a parent, a teacher or some other adult to fill you in on skepticism at some point. That lesson may never be delivered. Encouragement may never come your way. So take care of your own business. It's your life; live it wisely. Commit to skeptical thinking and try your best not to waste a single precious moment on lies and irrational dead-ends. Sure, you will make mistakes. You will believe a few stupid things along the way—it's inevitable, and even the best skeptics have. But try your best to keep those missteps to a minimum.

THE DAILY GRIND

Reading this book is a good start on the path to constructive skepticism, but it's only a start. You have to live it every day. It needs to become habitual, your instinctive way of thinking and reacting to the world around you. Do this, and you will spend the rest of your life in a much better place from which to thrive. Never forget, your brain belongs to you. It is the one asset that has immeasurable value. It's way more precious than a cool car or a flashy piece of jewelry. Make your brain work for you and not against you. Learn as much as you can about how it operates. This book will help, but it's your lifetime duty to keep learning. Understand and never forget that a brain weak on critical-thinking skills is more likely to trip you up *again and again* in life. Sadly, if your brain is not a top-notch reason machine, most of the time you won't even realize that you are a victim. You'll be falling when you think you're standing and shuffling backward when you think you are moving forward. Try your best to turn your brain into a formidable warrior, one capable of standing guard over you every waking minute of your life.

FANTASY AND FICTION ARE OKAY

Never think that being a good skeptic means one has to be antifantasy, hate fiction, and generally steer clear of anything that might be interpreted as make-believe. Skepticism is not opposed to things that do not exist, only to false advertising that says the fake is real. Fiction is great. Make-believe stories are wonderful. I would hate to live in a world that had no creative tales to brighten the nights and spark us to dream about what might lie over the horizon and beyond the next century. Storytelling and story-listening are big parts of being human. Made-up stories bind us together and help us to learn important things. The best fantasy and science-fiction writers seamlessly mix fiction with nonfiction to magnify ideas and lessons, making them memorable. Great stories drive us to imagine reality in new ways and then inspire us to make it so. Scary stories are mental theme-park rides that thrill us and tickle the deepest recesses of our brains to make us feel more alive.

Skepticism is not a threat or an obstacle to any of this. I'm a hard-core skeptic and I *love* science-fiction books and films. When I close my eyes, I see a time machine in my garage and terminator robots chasing people in the streets. I have visited a hundred planets and fought a thousand monsters—in my head. There is no contradiction here. I know many good skeptics who are neck-deep in the fantasy worlds of novels, films, or video games. It's only a concern when people confuse the fantasy for reality. As long as you can see where one world ends and the other begins, there is no problem.

SCIENCE VERSUS SUPERSTITION

We all have a choice to make about how we want to live our lives and how we want to think about all the things we will encounter over the years. Do we intend to lean toward the supernatural/paranormal stuff and relax our skepticism or do we want to go with science and keep it as real as possible? It's not always one or the other, of course. Many people jump back and forth between the two. I would argue, however, that it doesn't make sense to be science minded one day and superstitious the next. Pick a team. If the scientific method is the best way we have to figure out what is real, and if you agree that being aligned with truth and reality is wise, then why would you ever want to leave that path? I suggest you make up your mind which way will serve you best and then commit to it.

So which way is for you? Science or magic? Astronomy or astrology? Alternative medicine or medical science? Microbes or monsters? Crystals or quarks? Choosing between evidence-based science and faith-based magic shouldn't be too difficult. Don't overthink it. Just ask yourself the following: Which one does the best job of paying off in useful ways for you? Which one is more reliable? and Which one has the better record of success? Maybe the simple thought experiments below can help you to decide:

- You need to communicate with your grandmother urgently. But she lives two thousand miles away. What do you do? Do you close your eyes and begin chanting Grandma's name in the hopes of establishing a psychic connection? Or do you simply call her on a piece of technology produced by science known as a phone?
- You want to visit Hawaii for summer vacation. Do you attempt to travel there via astral projection or do you board a jet, another product of scientific thinking?
- Your best friend accidentally cuts her hand with a knife and is bleeding profusely. Do you rush her to a faith healer so that he can pray for the cut to close? Or do you rush her to a hospital, where doctors and nurses apply medical science to treat injuries?
- A young child loves space and dreams of discovering amazing things about the universe one day. For his birthday, you want to give him a gift that will help him toward that goal. Do you give him an astrology book or an astronomy book?

It should be obvious that science works. Even people who hate science and oppose it publicly and politically rely on it every day. I only wish it were obvious to everyone that paranormal and supernatural claims don't work. Science is reliable. It's not perfect, of course, but it's better than anything else we have. The long list of scientific achievements proves it. Just look around, for all the hype and noise, modern civilization certainly is not running on magic. It rests on a foundation of science and it is powered by science.

Please don't misread my words and think that I am suggesting science is the perfect and always-good counterbalance to irrational beliefs. Science is not a religion. It's not a moral system. Science can be bad, too, *very bad*. Science is also wrong much of the time. Scientists can be dumb and dishonest just like anyone else, as well. Try not to think of science as an institution or big organization like Congress or Microsoft. Science is best thought of as a tool. And, like most tools, it can be used to do something constructive or to whack somebody over the head. Science is a great way of thinking and discovering that helps us figure out much about the world and the universe. But as a collection of current facts and wisdom, it's not perfect—not even close—and to its credit, it doesn't claim to be. Always keep this in mind. One shouldn't believe in or revere "Science" in a way that might be described as worship or unwavering loyalty because we know that some significant portion of today's scientific knowledge is almost certainly wrong and will need to be corrected in the future.

Science is used to do bad things all the time. Science is the reason many of us can live longer and more comfortable lives than our ancestors did, but it's also the reason some countries have thousands of nuclear weapons and are capable of killing billions of people and destroying civilization in less than an hour. Science gives us a life-saving vaccine one day and a chemical weapon that melts your face the next day. Science can be used for good or bad, it's for us to decide. One thing can't be denied, however: It works. What about magical thinking?

How much good has superstition done for the world? The bad is easy to find; the good, not so much. Supernatural and paranormal beliefs have inspired and sanctioned tremendous abuse and exploitation over the centuries. From the horrors of human sacrifice and the routine terror of the Dark Ages to the trillions of dollars fleeced from weak skeptics by con artists today, magical thinking has always come with a huge price. But what about the positives? What good has it brought the world? Not as much as many claim, it turns out. It may be true that many people have found hope and reassurance in the warm embrace of the make-believe. But at what cost? And was this the only source of comfort available? I don't think so. Could many of those who clung to superstition not have turned to their family and neighbors for love and support instead? In many cases, superstition comes between human relationships. Relying on shamans and gurus who sold them empty air and called it magic might have felt comforting for many people in times of need, but perhaps they could have found as much or more elsewhere if they had tried. Certainly good skeptics have no problem hoping, finding inspiration, and enjoying the embrace of fellow humans during difficult times. It is a lie to say that a world dominated by truth and reality would necessarily be a world without hope and comfort. Many good skeptics live in that world as much as they can right now and are doing just fine.

IT'S NOT ABOUT INTELLIGENCE

A mistake many people make is to assume that thinking like a scientist in everyday life comes down to intelligence. Not only is that assumption wrong, it's a destructive misconception that probably causes many people to shy away from adopting a skeptical outlook. They aren't sure they are smart enough to think like a scientist, so they don't bother trying. The truth, however, is that all of us who fall well short of genius status can be very good skeptics. And the reverse of this is true as well: People who are extremely intelligent can be weak skeptics and fall for virtually every other nonsense claim that comes along. I've seen it with my own eyes. A brilliant or gifted brain doesn't do you any good if it's not used well and consistently. Ten soldiers with handheld rocket launchers might be the most lethal killers on the battlefield *potentially*, but if they aren't trained or are too scared to use their weapons in a timely and appropriate manner, one guy who has a crossbow, and the will and skill to use it, could take them all out.

Being smart, whatever that word means to you, doesn't automatically make someone a good skeptic. It can help, of course, but it's just not the most important factor when it comes to defending yourself against nonsense and lies. I once knew a bright guy who was convinced that there are large, technologically advanced, secret cities underground in various locations around the world and the people who live in them control everything on the surface. That may be a dumb idea, but he wasn't dumb. Many men and women with doctorates and master's degrees are weak skeptics who believe things that a good high-school-age skeptic can dismantle in seconds. My teenaged son is a far better skeptic than about 75 percent of the people I went to college with, including several straight-A engineering students. I once worked with a woman who was well educated and would probably be described as intelligent by most people. Smart as she may have been, however, she was absolutely convinced that a Russian girl had x-ray vision and could see inside of people with her bare eyes. My coworker read about the claim and swallowed it—hook, line, and sinker. If true, this extraordinary power could be easily verified with credible scientific testing. But it never has been confirmed, so it makes no sense to believe it. My former colleague was typical of the highly educated weak skeptic that is all too common today. She had a sharp mind and achieved more academically than most, but in school she apparently was taught only *what to think* and not *how to think*.

History offers numerous examples of very smart people who believed very goofy things. Today Isaac Newton is considered by many to be no less than the greatest scientist who ever lived. It's tough to argue against him. Toiling in the 1600s, Newton managed to invent calculus and figure out how stars and planets move about in the universe. But he didn't spend all his time on the work that would make him so famous. This undeniable genius also devoted much of his time and energy to the Christian apocalypse claim. He seriously thought he could calculate the exact date of a supernatural doomsday and did come up with one. (You might not want to make any plans for the year 2060.) Newton was a great mathematician, but not a good skeptic. Had he been, he might have recognized early on that his time would be better spent on other things. We can cut him slack given the time he lived in, of course, but what other great things might the brain of Isaac Newton have contributed to science if it had been protected by the force field of skepticism? Fast-forward to today, and one wonders what contributions to science and the world won't happen because some of the twenty-first century's brightest minds are distracted or derailed by bad ideas that skeptical thinking could easily brush aside if applied?

Why do you suppose it is that the world's professional astronomers spend virtually no time on astrology and UFOs? The reason is that they are good enough skeptics to know that their efforts will be better spent elsewhere because nothing is likely to come from trying to work out the influence of Jupiter on the love lives of people born in September or determining if the alien crew survived the Roswell UFO crash or died on impact. There is an invaluable life lesson in this for you. Everyone loves to say that hard work is the key to success. But few explain that hard work is only part of it. What you direct all that effort at matters a lot, too. You can work twelve-hour days, seven days a week, year after year, but if it's all toward proving that some houses are haunted, you are probably wasting your time because skeptics have shown over and over that there is no good evidence and there are no compelling reasons to believe it. It's safe to assume, for example, that the hardest working psychic in history has contributed less to the world's understanding of the human brain than, say, one lecture by one random psychology professor at any university.

This is not to suggest that some things should be off limits for inquiring minds. *Everything* is worth a thought or taking a peek at because one never knows what discoveries lie just beyond the reach of current understanding. I'm not for banning ghost hunters or making UFO-watching illegal. People should be free to think whatever they want and to follow their passions where they lead. However, it's in one's own interests to be reasonable and realistic when deciding what to pursue and devote limited time and energy to. For example, the Yeti, or abominable snowman, might be real—it's possible—so I suppose you could spend a career looking for it. But if discovering an unknown species is your dream, then it makes much more sense to look for new arthropods or microbes in deep caves or in the Pacific abyss than it does to hunt a giant primate in Tibet. A basic skeptical analysis of the Yeti claim and a review of all available evidence make it clear that it is among the longest of long shots, to the say the least. However, based on past efforts, it's hard to imagine *not* finding a few previously unknown arthropods and microbes if one looks in the environments they are likely to be in.

SKEPTICISM TAKES ON ALL COMERS

I hope that it is encouraging, for young people in particular, to recognize that one doesn't have to read a thousand books or spend ten years in graduate school in order to become a razor-sharp skeptic who is not easily fooled by false claims and inane ideas. Learning relevant facts and having a general understanding of how the universe works can help explain away many bogus beliefs, of course. But far more important than reading, research, and schooling is committing to the skeptical attitude in the first place. It's about having the will to ask the necessary questions and the courage to walk away if the answers don't measure up. It's about withholding belief until you know. The process of skeptical thinking works very well *even when you don't know much or anything at all about the particular claim.* For example, there are so many weird alternative medicine products and weight-loss gadgets on the market today that I can't keep up with them. But it doesn't matter. If someone approaches me later today with something I've never heard of—let's call it the "Beejeepers Pill" that is "guaranteed to prevent and cure cancer"—I already know what to ask and what to look out for because I'm a good skeptic. I don't have to become an expert on the Beejeepers Pill and understand everything about its chemical makeup and specific history in order to approach the claim in a sensible manner. Here is how it would most likely go down from my side of the conversation alone:

- *Prevent and cure cancer? Wow, that must be an amazing pill you have there. By the way, I'm curious, what are your credentials in the medical field?*
- *How exciting this must be for you! If your pill works, it means you will be responsible for saving millions of lives each year. You will become one of the richest and most famous people in the world, one of the greatest scientists in history. A Nobel Prize in medicine is guaranteed. You must have been so excited when the test results came in.*
- *You did do scientific, double-blind tests on the pill, right?* [Double-blind test, or trial, means neither the researchers nor the patients know which test subjects get the real Beejeepers Pill and which ones get sugar pills. This helps to prevent the results from being contaminated by bias.] *And surely you published the results in a respected medical journal, right?*
- *But I don't understand; why haven't you done proper testing of the pill or published anything in a journal? Don't you think you need to do that before you start selling it to people? How will anyone know if it really works or if it's just more medical quackery?*

- *Hold on, my friend. You are changing the subject. Why are you suddenly telling me random stories about people who you say took this pill and got better? Of course you know that mere stories about a few people don't prove anything, right? Even the worst, most ineffective rip-off medicines in the world have stories like that to back them up.*
- *Hey, wait. Why are you leaving? Well, okay, it's been nice talking to you. Please get back to me when you can prove that your pill works.*

The ability of a good skeptic to mount a respectable challenge to virtually any unusual claim that comes along is amazing when you think about it. One does not need to be an expert on the particular subject to at least put up a sound skeptical defense against empty claims and unlikely beliefs. I'm just one guy, but I do pretty well in discussions/debates with the endless parade of people who try to convince me that weird things are true. I don't have to be smarter than them or even more informed on the specific topic in question. I just have to be a good skeptic. I am frequently a guest on call-in radio shows, for example, and I never know what will come up next. Callers have asked me about everything from UFOs and the Moon-landing hoax to Laetrile and the lost continent of Atlantis. But I feel that I've managed to field every challenge to date very well. A strong skeptic core enables one to take on all comers and do very well in the wild, dark forest of extraordinary and unproven claims. Although knowledge is important, skepticism is not a body of knowledge. It's a way of thinking. It's less about something you know and more about something you do.

NEVER GIVE UP! NEVER SURRENDER!

Becoming a good skeptic is about making a promise to yourself that you will not be an easy victim. Sometimes you really have to view this as war, not against people but against lame ideas. Tell yourself that you won't go down without a fight when people try to convince you to believe in outlandish things that they can't back up with very good evidence. Some people who try to push irrational beliefs on others do it with a strange enthusiasm. Some can be relentless about it. They are like starved zombies banging on the door in search of their next meal. Their goal is to bring you into their undead legion of the unthinking. Don't let it happen. Don't let them in. Better they and their vacuous claims splat against your force field. Nudge these people and their claims into the meat grinder of the scientific process and see how well they hold up. Ask good questions and demand plenty of evidence. Stand up for yourself always. Think for yourself always. And whatever you do, don't let them in!

THE GREEN TROLL IN A RED DRESS

When you are confronted with a weird idea, it is important to keep in mind a simple statement popularized by the late astronomer Carl Sagan, "Extraordinary claims demand extraordinary evidence." Remembering this when it counts can save you time, money, embarrassment, and possibly even your life. There is great power in these five words, enough to slay dragons and repel the crooks and crazies.

To put it into action, the first thing we need to do is to assess the size of the claim so that we can assess the amount of evidence that we need to accept it. No, put away the scales and tape measure. A general estimate is all that is needed. For this, "a little" or "a lot" are fine units of measure in most cases. It's about seeking a rough balance, that's all. We don't demand mountains of evidence for every claim anyone makes about anything, of course. That would be overkill and would waste too much time. For example, if for no particular reason I tell you that I saw a blue car driving on the road near my house yesterday, you would be right to go ahead and believe me. Seeing a blue car is not an extraordinary claim. There are millions of blue cars. It's reasonable and likely to be true that I saw one. And so what if I did? It's not a big deal. It's not like I'm asking you to attend my weekly blue-car-sighting meetings and make a monetary donation on behalf of the blue-car fund of America. You can probably go ahead and trust me on this one. But what if I make a very different claim? What if I say that yesterday I saw a winged troll with green skin wearing a spectacular red dress and playing "God Bless America" on a jewel-encrusted harmonica while flying around in my backyard? Are you going to accept my claim and believe it happened? What if I also tell you that this particular troll is short on cash and needs $100 per week—which I have been authorized to collect for him? I hope that you wouldn't trust this story without more effort to prove its veracity on my part. It's an extraordinary claim, so it demands *extraordinary evidence*, right?

You don't have to disprove the troll story. That might be impossible. We don't know with absolute certainty that winged, green-skinned trolls do not exist. Maybe they do, and if they do exist, we don't know that they don't need money. We can't even say for sure whether or not they look great dressed in red. But that's okay. Doubt is enough. Being a good skeptic doesn't mean you have to know everything. It certainly doesn't mean you have to disprove every claim believers make.

The burden of proof is on me for making the claim in the first place. For a whopper of a tale like this one, I should have to produce tons of very good evidence if I want sensible people to believe me. Sadly, of course, if I wanted to and was willing to work at it a bit, I have no doubt that I could find a few weak skeptics somewhere who would believe in my green troll wearing the hot dress—no evidence required. Do you doubt it?

IT'S ALWAYS A GOOD TIME FOR SKEPTICISM

If someone were to tell you that he has a time machine, an extraordinary claim if ever there was one, it would only make sense to ask him to prove it. No fancy talk, no hype, no interesting stories, and no intriguing promises, just a lot of good, solid evidence. But first, what is this thing called "evidence" that skeptics like me are always talking about? It's simply stuff that helps make the case, things that support a claim. It can come in many forms, from photographs to a strand of hair to an eyewitness account. Not all evidence is equal, however. Some evidence is so weak, it may as well not be there; and some is so strong that it qualifies as proof. One thing is consistent, however: When it comes to extraordinary claims, the more evidence, the better.

For the particular claim about a time machine, you might ask your buddy to bring a live baby triceratops back from the past or go into the future a few centuries and fetch the cures for cancer and malaria. But even those efforts would not be perfect proof. Maybe he obtained a live triceratops from a successful cloning effort using DNA harvested out of fossilized bones. Maybe he already had the cures for cancer and malaria here in our time and lied about getting them from future. But this sort of evidence would be so difficult to come by without a time machine that it would make the claim much more credible and well worth investigating further. However, what if the alleged time traveler brings back a message from the people of twenty-fourth century that warns us to protect the environment and learn to live in peace with one another? *Yawn.* Not good evidence. Even worse, he might argue that he shouldn't have to prove something so special and wonderful. It's rude to even ask for evidence, he might say. You should just believe what he says—and, by the way, he needs some money to maintain his time machine and fund more trips. This is the point at which good skeptics smile and walk away.

BELIEVERS SAY THE DARNDEST THINGS

It is always important to keep the big picture in mind when trying to think critically about extraordinary claims. Don't allow a person trying to sell you a freaky belief to slip in a false premise on you in order to make the conclusion seem more reasonable. A typical line goes like this: "As everyone knows, an extraordinary number of ships and planes vanish mysteriously and inexplicably in the area known as the Bermuda Triangle. Therefore there must be substantial paranormal activity occurring there."

Never forget that *everything* a person making a strange claim says should be analyzed and challenged by your skeptical brain. Forget for the moment that supernatural or paranormal powers may or may not be at work in the Bermuda Triangle. The first thing to question would be the claim that an "extraordinary number" of ships and planes go missing without explanation in the area called the Bermuda Triangle. One might check with the people who would pay close attention to such a trend and know more than anyone about it—like the US Navy and the US Coast Guard. (I did; more on that in chapter 3.) You might also question the very existence of something called the "Bermuda Triangle." Just because people talk about it or it shows up in some slick TV "documentary" doesn't mean it is necessarily real. Did you know that no such triangular region shows up on any official government or military maps—or on any other serious maps made by credible map-making companies such as Rand McNally or National Geographic? No group of professional and qualified cartographers or scientists has ever agreed on any boundaries for this "place." Just saying something doesn't make it so.

Belief peddlers also like to throw in something that is true but does not prove their original claim just because it sounds good and might make them seem more credible. This verbal sleight of hand is common, so watch out for it. If "X" is obviously true and someone attaches it to "Y" in order to come up with "Z" as a conclusion, then we always need to make sure that "X" and "Y" really have something to do with one another in the first place. Like I tell my kids all the time, one tiny, careless mistake in a long math equation makes the final answer wrong. You can't afford to be sloppy and rush over any of the steps or variables, both in math and in skepticism.

Many people who push paranormal and supernatural claims are also in the habit of distorting reality to try to make their case. Don't let them. When someone says that "everyone else" believes her claim, pause and think about it. Is that true? *Everyone?* I doubt it. People are in the habit of saying, "they proved" and "they discovered." You need to know who *they* is before you allow yourself to be impressed by *them*.

Don't be misled by inconsistent skeptics. Many people who can't resist spending their money on medical quackery might happen to reject psychic readings, for example, and vice versa. I met a person who laughs at the Moon-landing hoax claim but does believe that the US government has been reverse engineering alien spaceships at Area 51. Always listen carefully to the entire sales pitch and remember that no one person is always right about everything.

Finally, don't let anyone inflate one wild claim with other wild claims in an attempt to overwhelm you with stories, data, or arguments. Believers often inject secondary unproven claims into the mix in order to make it seem as if they are building a strong case when they are not. Call them on it. Pause the conversation and ask about each claim. For example, if someone tried to tell me that invisible rays from the Moon have been making a significant impact on how mediums talk to the dead, my response would be: *We'll get to mediums soon enough, but first tell me about these Moon rays. What are they? Does NASA know about them? Why not?*

If someone said that psychics have determined that horoscopes are 67 percent more accurate on Thursdays than on Mondays, I first would ask for an explanation of why psychics should be viewed as a reliable source about the claim before I entertained anything about horoscopes.

EVIDENCE? WE DON'T NEED NO STINKING EVIDENCE!

Consider it a warning sign whenever someone downplays the need for evidence and insists that it's unnecessary. "Faith is enough," is a common claim. When you hear this, immediately go to DEFCON 2, signal red alert, and raise shields. When they say, "Just trust me," it's time to throttle up the skepticism and man your battle stations. Relying on faith might sound sweet and sensible to some ears because it's heavily promoted as something positive and reliable. But the truth is that faith is really nothing more than pretending to know something that you don't know. It's worse than bad thinking; it's antithinking. Faith as a means to figuring things out, coming to a conclusion, and justifying a position on something is the opposite of thinking like a scientist. It's a terrible way to make your way through life because it can lead you to believe just about anything.

The crucial problem for those who say faith is an acceptable way of approaching unusual claims is figuring out where to draw the line. When exactly are we supposed to rely on faith instead of thinking? If I decide to have faith and accept the claims of some astrologer or doomsday prophet, should I also have faith and accept claims that gargoyles and fairies are real? Should I use faith to believe in mediums who say they talk to dead people, even though they can't prove it? What's the difference? If blind trust works for one thing, why not for everything?

"I COME IN PEACE"

When discussing an extraordinary claim with a believer, I recommend being friendly about it. It's usually better for everyone. Just don't be so nice that you let yourself become a punching bag. You have to stand up for your brain and you have to work to keep the conversation somewhere inside the general vicinity of honesty and reality. Many people who believe weird things veer off into fantasyland every chance they get and try to take you with them. They usually seem to do this without even realizing it, so I tend to be forgiving about it. No matter how strange a conversation may become, however, I never assume people are lying or are hopelessly stupid. Maybe they don't realize how badly they are abusing logic and torturing common sense. Maybe they haven't given the subject much thought and are just winging it. Maybe they are hypercompetitive and would rather spew crazy talk than risk going silent and "losing" by default. Regardless of why they do it, however, it is the responsibility of a good skeptic to keep things as sensible as possible.

Remember that you're not rude for doing nothing more than asking people to explain, defend or prove a belief that they are trying to talk you into accepting. I would never go to the home of someone who believed in an imminent Judgment Day, knock on the door, and demand proof that the world is about to expire. But if people come to my door and declare that these are the final days, then it's open season on their claim. I can ask questions and demand all the evidence I want. You are never wrong for being skeptical of an unusual or important claim, and you are never a jerk for challenging a claim when someone dumps it in your lap. Just because others think it's the greatest thing ever doesn't mean you have to suspend your thinking. Besides, if their belief can't hold up to a little scrutiny, then you are probably doing them a favor by at least presenting them with the opportunity to reconsider their conclusions. However, if a belief is doing harm to someone, then that's plenty of justification to challenge it anytime, regardless of context. It sometimes helps to remind believers that true claims have nothing to fear from skepticism. If believers refuse to think critically about their claims, then call them on it. *Why are you reluctant to challenge a claim that you say is so important and obviously true? What are you afraid of?*

One can easily avoid being too harsh to sensitive ears by softening the delivery of a skeptical challenge. You don't even have to "tell" them anything. Simply asking questions works wonders. Blasting a believer with the blunt force of facts, declarations, and accusations can feel like a mean-spirited attack. Questions tend to feel much gentler. Here are some hypothetical exchanges between a believer and a skeptic to help illustrate the point:

- **Believer: "It's amazing how accurate my horoscope was today. It hit right on the mark!"**

  Skeptic: *But was the horoscope vague? If it was specific, couldn't it just be a coincidence? After all, how many times has your horoscope been wrong? A few hits in the context of thousands or millions of misses don't really prove anything, right? Oh, and how does astrology work, anyway? Is it by gravity or what? If so, exactly how does the gravity of a planet or star influence a person's career and love life?*

- **"Wow! That psychic knew so much about me."**

  Are you sure the psychic really knew important and unique things about you? Or is it possible that the psychic threw out a bunch of edu-

cated guesses and then read your verbal and physical reactions to them? Is it possible that you are remembering a few of the guesses that seemed to be correct while forgetting the many more guesses that were way off the mark?

- **"It's a miracle that I survived cancer."**

  I'm glad you are okay, but are you sure this was something that we should call a miracle in the sense that it was a supernatural event? Was a doctor treating your disease? Were you taking medication, getting chemotherapy? If so, why are you saying it was a miracle instead of medical science that saved you? Cancer is often deadly, of course, but millions of people do survive it every year. Is every case of a person beating cancer a miracle?

- **"I had a dream last night and it came true."**

  Worldwide, how many dreams do people have every year? A lot, right? It's got to be in the hundreds of billions. So doesn't it seem likely that out of all these dreams, many thousands of them are bound to match up with future events, just by chance? Also, is it possible that you might have been influenced by the dream and subconsciously or consciously done things to make it come true?

- **"I saw a ghost last night."**

  How do you know that it was a ghost that you saw? How can you be sure that it wasn't some natural thing that your brain misinterpreted as supernatural? How do you know it wasn't your imagination? It's common for smart, normal, sane, and honest people to "see" things that aren't really there. It happens all the time. Given the way our brains work, don't you agree that it's not reasonable to completely trust eyewitness accounts when it comes to important or unusual things such as seeing a murder or encountering a ghost? What if I told you I saw Elvis at a shopping mall yesterday? Would you believe me? Why not?

- **"A teaspoon of natural extract of fried gopher nipples cured my flu."**

  Mmmm, sounds yummy, but how can you be sure that it cured your flu? Maybe it was just a combination of time and your body's own defenses that did it. I've been sick with the flu and it went away, despite the fact that I never took fried gopher nipples for it or any other alternative medicine.

WRAP IT UP

Those who try to convince you that skepticism is negative, pessimistic, or destructive are either profoundly clueless on the matter, or they are being deceptive because they hope to sell you something that you don't need. Skepticism is positive, optimistic, and constructive. It's the invaluable defensive weapon that you want and need every day of your life. The only people who should not be good skeptics are those who have a deep desire to be a victim, to be foolish, to waste time and money, and perhaps to risk their life for nothing. Three things make it abundantly clear why it is wise to be a good skeptic:

1. There are dishonest people in the world who are ready to trick you out of your money by getting you to believe claims that are very likely to be untrue because the evidence is weak or nonexistent.
2. Thanks to science, we now understand more than enough about how the human brain works to know that we have to be very careful when trying to determine what is real and what is not, what is true and what is false. If you are human, you can and will be fooled by your own brain in various ways many times throughout your life. Being a good skeptic is the only way to reduce these missteps and minimize harm from them.
3. Smart doesn't mean safe. No matter how brilliant you may be, don't fall into the trap of assuming that intelligence alone will keep you safe from nonsense beliefs and clever frauds. One of the most important messages in this book is that we are all vulnerable. Arrogance can lead any one of us right into the open arms of those who peddle irrational junk. We must be on guard *all the time*. Many wacko claims are customized and packaged in ways that make them seem reasonable to reasonable people. Some may be crafted precisely to appeal to people who are exceptionally smart. Even worse, crazy claims and bogus beliefs don't always come to us from creepy strangers. They are often delivered to us by trusted loved ones, people we admire, and respected authority figures. Good skeptics know to always be ready to think, regardless of where they are or whom they are listening to.

You are free to believe before you think if you wish. But just know that taking such a stance in life puts you at risk. You may end up as one more human piñata, a fat juicy target of opportunity for all those eager to fill your head with their baloney or empty your pockets in exchange for lies. They will come for you, again and again, and they will hammer you relentlessly with delusions and deceptions. The force field of skepticism is the only thing between you and them. Keep it running at full power.

GOOD THINKING!

- Weak skepticism is perhaps the greatest unrecognized global crisis of all. Every day, people waste time, throw away money, suffer, and even die because they failed to think like a scientist.
- The quality, strength, and consistency of your skeptical thinking will likely have a direct impact on your safety, success, and quality of life. Doing your best to stay free and clear from bad ideas and bad people who would take advantage of you should be a top priority.
- Anyone can think like a scientist. It doesn't matter who you are, how old you are, or how smart you are. The key questions are: (1) Do you have the courage? and (2) Do you care enough about your life to protect it from the nonsense and the madness? Skepticism is the path toward a lean and efficient life, one that affords more time and energy for romance, family, friends, fun, and creativity. Think and get the most out of your life. Think and be fully human.

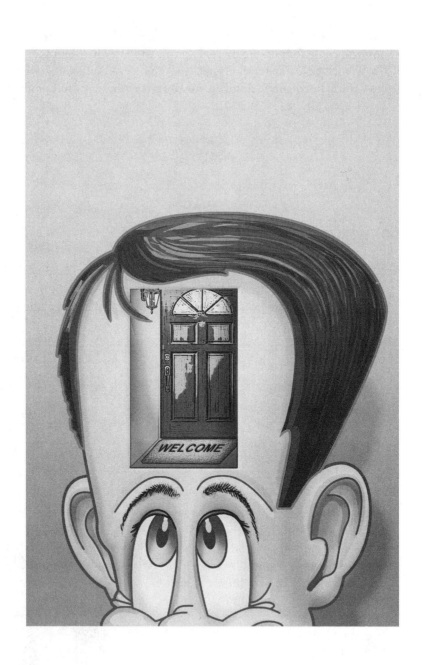

Chapter 2

PAY A VISIT TO THE STRANGE THING
THAT LIVES INSIDE YOUR HEAD

Once you understand something about how our brains operate and how easily they lead us to see, hear, and feel things that aren't there, believe in things that don't exist, and think things that make no sense, you might never again find the courage to walk out of your front door. I have been paying close attention to the work of psychologists and brain scientists for many years, and at times their discoveries have left me reeling. Unfortunately, most people know little or nothing about how the brain operates so they make incorrect assumptions about its reliability. The brutal truth is that human brains do a poor job of separating truth from fiction. This leads to many false beliefs. Therefore, it's only wise to make at least a minimal effort to understand the brain and be on guard against its deceptive ways.

Many of the weird yet routine processes of the human brain are not only extremely interesting but have direct relevance to the safety and quality of our lives as well. It's like we are all sharing our bodies with another person, somebody called "Brain" who we barely know. This odd roommate is eccentric, is secretive, and doesn't ask for our consent when it goes about most of its business. But it does mean well and performs important work that we literally could not live without. So before we address the problems in our heads, let's make sure we appreciate the human brain for all the good it does.

The brain processes all the tasting, smelling, hearing, seeing, and touching that we do throughout our lives. It regulates breathing and blood circulation. It stands guard over us, ready to alert us to injuries and other problems via a messaging system called pain. And, of course, it's the place where all our thoughts and memories live. It is the beginning and end of us all. No brain, no you.

Think of what human brains have accomplished to date. They imagined and then made wood and stone tools that allowed our ancestors to rise above challenging environments and overcome stronger, faster predators. They created music, humor, and countless fantastic stories and ideas, some of which reach beyond even the limits of our universe. About 80 to 100 *billion* cells called neurons make up this amazing three-pound organ. These cells are connected to each other by some 100 *trillion* little structures called synapses. Electrical and chemical messages fly around in the brain constantly, making it a nonstop hub of action. Even during sleep, the brain keeps working for us.

The problem is that our brains go about many of their duties in very strange ways that most people are unaware of. But you had better clue in fast, because having at least a basic understanding of what that strange thing inside your head is up to is absolutely necessary to being a good skeptic. People who have no idea how human vision and memory work, for example, are far more likely to see an alien spaceship in the sky or a ghost down the hall and then confidently remember the event in great detail and with confidence. Because of their lack of awareness, it may never occur to them that natural processes of the brain present a far more reasonable explanation for what they thought they saw and what they think they remember. Those who understand that our brains don't play fair when it comes to assessing arguments and evidence are far less likely to spend years of their lives clinging to lies, lame ideas, and bad beliefs. Without some basic knowledge of how stories seduce our brains, you might run right by a pile of solid scientific evidence in order to snuggle up with one tall tale.

I thought I was well informed about cognitive biases, memory flaws, hallucinations, and so on. But while researching my book *50 Popular Beliefs That People Think Are True*, I was repeatedly shocked by new things I learned from reviewing scientific research and talking to experts. The inside of a human skull is a far freakier environment than I had ever imagined. That realization helped to inspire this book. I kept thinking: *People need to know about this!* Science is revealing more and more every day about how our brains work and how they deceive us both in unusual circumstances as well as in routine, everyday life. It can be a rude awakening to realize that you are walking around in a reality that your brain *created* for you, a "reality" that can never be trusted completely.

REMEMBER NOT TO TRUST YOUR MEMORY

Did you know that you can't trust even your most precious memories? They may come to you in great detail and feel 100 percent accurate, but it doesn't matter. They easily could be partial or total lies that your brain is telling you. Really, the personal past that your brain is supposed to be keeping safe for you is not what you think it is. Your memories are pieces and batches of information that your brain cobbles together and serves up to you, not to present the past as accurately as possible, but to provide you with information that you will likely find to be useful in the present. Functional value, not accuracy, is the priority. Your brain is like some power-crazed CIA desk jockey who feeds you memories on a need-to-know basis only. Daniel Schacter, a Harvard memory researcher, says that when the brain remembers, it does so in way that is similar to how an archaeologist reconstructs a past scene relying on an artifact here, an artifact there.[1] The end result might be informative and useful, but don't expect it to be perfect. This is important because those who don't know anything about how memory works already have one foot in fantasyland. Most people believe that our memory operates in a way that is similar to a video camera. They think that the sights, sounds, and feelings of our experiences are recorded on something like a hard drive in their heads. Totally wrong. When you remember your past, you don't get to watch an accurately recorded replay.

Even knowing what I know about human memory, however, I continue to live as though my memories are reliable and accurate. One has to in order to function in daily life. Most of the time it all works out fine, which is why the human brain does it this way. But it can be tough sometimes because I've peeked behind the curtain. I've seen the Wizard. I know what's really going on in my head. No longer can I be absolutely sure about all the places I've been and the things I've done. For example, one special day long ago I found the courage to kiss Kimberly, the most beautiful girl in my third-grade class. Or did I?

Did I really start that glorious day as I remember, by giving her the note I had spent an hour crafting the night before, the one that asked her to be my girlfriend? I can see it in my head right now. There was just one line: "Do you love me?" There were three boxes below it for her to check *yes*, *no*, or *maybe*. After the final bell that day, did I really man up, as I recall, and lean in with my little eight year-old body to execute the most romantic kiss on the cheek in the history of all elementary

schools? I can see the replay in mind, clear and in great detail. I'm confident that I remember it correctly. But did it really happen that way?
Because of the way our brains work, I can't be sure, no matter how
confidently my memories assure me that it did. Who knows? Maybe it
did happen, but it was in fourth grade rather than third. Maybe it was
Kimberly I dreamed of kissing but it was another girl whom I kissed.
Or worse, maybe I missed Kimberly's cheek and kissed a goat by accident while on a school field trip to a farm. Maybe my classmates ridi-
culed me for it on the long bus ride back to school. I don't remember
anything like that happening. But perhaps that's because my brain
edited and embellished a bitter memory to help me cope—all without
consulting me first, of course. It's possible because, like it or not, that's
how human brains work when it comes to memory. Was Kimberly a
goat? I hope not, but short of photographic evidence of the big kiss, or
documentation from an official animal-endangerment investigation, I'll
never know for sure. Don't laugh at me—for all you know, there may be
a goat or two in your past since all human brains operate on a foundation of creative deception when it comes to memory.

To try to make it easier for people to understand how memory really
works, I describe it like this: Imagine a very tiny old man sitting by a
very tiny campfire somewhere inside your head. He's wearing a worn
and raggedy hat and has a long, scruffy, gray beard. He looks a lot like
one of those old California gold prospectors from the 1800s. He can be
grumpy and uncooperative at times, but he's the keeper of your memories and you are stuck with him. When you want or need to remember
something from your past, you have to go through the old codger. Let's
say you want to recall that time when you scored the winning goal
in a middle-school soccer match. You have to tap the old coot on the
shoulder and ask him to tell you about it. He usually responds with
something. But he doesn't read from a faithfully recorded transcript,
doesn't review a comprehensive photo archive to create an accurate
timeline, and doesn't double-check his facts before speaking. He definitely doesn't play a video recording of the game for you. Typically,
he just launches into a tale about your glorious goal that won the big
game. He throws up some images for you, so it's kind of like a lecture or
slideshow. Nice and useful, perhaps, but definitely not reliable.

Your memories are *stories* about your past. Without informing you,
the old man might take it upon himself to leave out some parts of the
story just because *he* feels they are unimportant. He also might get

confused and accidentally toss in a few scenes from that horror film you watched last year. He's good at that. He can blend the past you are recalling with another past or an entirely made-up event in seamless ways that you probably won't catch. Also, as with most stories that are told and retold, the key elements, names, and imagery all change over time. With each telling, the facts and people can easily get shuffled or lost entirely. That's the reality; your memories depend on the whims of an inconsistent and somewhat flakey storyteller. Don't get me wrong, I'm not condemning the old man. He does all this to try to be helpful to you. And it usually works just fine. It's the best way most of the time because we don't want or need to remember everything we see, hear, and experience in life. It's too much information, and a brain can only store and utilize so much. The key point to keep in mind is that when it comes to remembering very important or unusual things—such as a murder, an alien abduction, or a ghost sighting—the old man might not get it right, so we can't trust memory alone.

I'LL NEVER FORGET THAT DAY! (*YES, YOU WILL*)

Contrary to popular belief, even our memories of dramatic and "unforgettable" events are not reliable. This is hard for many of us to believe because we tend to assume that a big, spectacular moment in our lives will be "burned" into our minds forever. Sorry, but that's just not true. Researchers have shown this repeatedly by doing things like asking people to write down where they were and who they were with when something big went down, such as the 9/11 attacks. They did this shortly after the event. Then, when asked the same questions again years later, many would confidently remember being with people and in places during the event that contradicted their earlier and presumably more accurate memories of the day.[2]

Perhaps the biggest memory problem that good skeptics need to be aware of is that recollections of real events can be altered or contaminated with false memories and imagination. False memories can even be implanted by other people. This can be done fairly easily, much to the horror of people who care about truth and reality. Memory researchers have shown that doing nothing more than mentioning a name or showing an image of something before asking someone to recall an event can significantly change how he remembers that event.

For example, if I was going to test your recall of your first day of high school, I might casually chat with you beforehand and slip in a few comments about cold weather and snow. Then, during questioning about your first day in high school, I would ask: "Was it hot or cold that day?" Chances are, you would answer "cold," even if it had been hot. Yes, it can be just that easy to change a person's memory.

Even more strange, it is possible to implant a completely fabricated memory into someone's brain, not just finesse an existing memory but create an entire event. Again, this is not as difficult to do as you might think. By simply suggesting to someone that she was at a particular family picnic or party years ago, her brain might instantly conjure up a convincing memory of all the fun she had that day—even though she never made it to that particular event. People can end up with partially or totally manufactured memories inspired by films, novels, or stories they hear, too. The cold hard truth is that our memories are highly vulnerable to being altered by suggestions or just plain made up by the brain. You can't trust them completely. By the way, it's important to add that this has nothing to do with intelligence or mental-health problems. It's just how normal, healthy brains work. So, this is your memory and don't forget it. Now you know that there are very good reasons to be skeptical when people say they remember extraordinarily weird things. They may be remembering accurately, but how can you know for sure? You have to ask for additional evidence. Unfortunately, most people in the world haven't heard about what scientific research has revealed about human memory, so they are too quick to trust the memories of other people as well as their own memories. Fortunately, you know better now. It doesn't make sense for us to believe unconditionally when someone claims to have witnessed a paranormal or supernatural event yesterday, last week, or last year. Even if someone happens to be the smartest, most trustworthy person in town, we would still need good evidence to confirm that her memory is an accurate account of what happened.

THE DECEPTIVE BRAIN

Remembering the past accurately is not our only problem. Far from it. We aren't very good at thinking straight about the present, either. We all have many natural biases that have stealthy ways of distorting our thinking. What we may imagine is a reliable and logical brain is more like a three-ring circus of wacky thinking. For example, did you know that we instinctively notice and remember evidence that supports our beliefs while simultaneously ignoring and forgetting evidence that contradicts our beliefs? It gets worse. We can look at a scene before us, convince ourselves that we are paying close attention to it, and then fail to notice important, unusual, or unexpected things present, even though they are right in front of our open eyes. Yes, just because you may be staring at something doesn't necessarily mean you see it. Our brains also compulsively "connect dots" in order to create meaningful images out of visual input. This can be a useful ability in some situations, but in others it becomes a mental trap that causes us to see and believe in things that aren't there. These are only a few of the many problems we face when trying to figure out what's going on around us. You can go through life never knowing or caring about them, but why would you want to? Don't you want to give yourself the best chance possible to avoid mistakes in how you think and how you perceive the world around you?

After years of researching skeptic topics and the natural workings of the human brain, I have a lot of empathy for people who believe weird things that almost certainly are not true. It's so common to the human condition to be snared by irrational beliefs that I feel very fortunate not to be running around somewhere right now searching for fairies, preparing for the mothership to land, or giving away half of my income to some psychic or faith healer. The good news for you is that just being aware of how your brain goes about its business greatly improves your chances of keeping both feet planted in reality. When you realize that your brain is playing its own game by its own rules, then you know better than to trust it unconditionally and you are better prepared to take on the world.

FORGET EYES—IT'S THE BRAIN THAT SEES

Let's take a look at vision. Have you heard the old saying, "seeing is believing"? Well, it's often a case of *believing is seeing*. It is well known by researchers that what we think we see can be strongly influenced by images and ideas we have been exposed to previously as well as our own thoughts and imagination. This probably explains why it's the people who already believe in ghosts or UFOs who keep seeing ghosts or UFOs, and why so few nonbelievers do. Seeing things that are not there can happen to anyone because the human brain *constructs* and *interprets* the visual reality that is around it. What we see is something the brain has produced for us *based on* input it received via the eyes. It's never a 100 percent true and complete reflection of what our eyes are pointed at. For this reason, we can't always be sure about what we think we see. Yes, that might be an angel that you see up ahead. Or your brain could be showing you an angel that it has mistakenly constructed out of a bush or some other object.

Construct and interpret reality? It sounds crazy when you think about it. We don't really "see" the things we look at? How can this be? Most people probably assume that the brain simply shows us or somehow faithfully relays whatever images come in through the eyes. But that's just not how it works. What actually happens is that light patterns enter the eyes and electrical impulses are sent along optic nerves to the brain. Then the brain *translates* these impulses into visual information that you "see" in your head. Your brain doesn't reflect or replay the scenery around you like a mirror or a camera and monitor would. It provides you with its own highly edited and customized *sketch* of the scene. Your brain gives you a *version* of what you look at. It's as if your brain comes up with something like a Hollywood movie production that is loosely based on what is really going around you. You are not watching a video feed; you are watching a docudrama. The brain takes the liberty of leaving out what it assumes are unimportant details in the scene before your eyes. Just like memory, this is not necessarily bad most of the time. In fact, it's necessary in order to avoid information overload. You don't need to see every leaf in every tree and every blade of grass in full detail when you look around a park. That would be way too much data. It would clutter your thoughts and make you less efficient, if not incapacitate you. What you need in order to walk through a park and function well is to have a general picture of your surroundings, so that's what your brain gives you. If you need more detail, then your eyes and brain zoom in and focus on a single leaf or an individual blade of grass.

It gets weirder. Not only do our brains leave out a tremendous amount of detail, they also routinely fill in gaps in our vision with images that you can't possibly "see" or that maybe don't even exist in reality at all. Your eyes might not be able to track a fast-moving object, for example, so your brain will sometimes conjure it up and show it to you anyway, figuring that it might be useful to you to see a projected version of reality. The brain also fills in missing elements that "should be there" in static scenes because, again, it can help us to navigate our way through the environment. Magicians have known about this for many years. Even if they don't understand or care about the science behind it, they take full advantage of the way our vision works when they do their sleight-of-hand coin tricks, for example. Again, our brains don't do any of this for a gag or to make fools of us. They do it because it is the most effective and efficient way to function in life most of the time.

In addition to filling in missing images, our brains also find patterns or connect the dots when we look around. They do this automatically and do it very well. It helps us to see things that otherwise might be difficult or impossible to recognize. It's probably one of the main reasons you and I are alive right now. Like many other animals, our prehistoric ancestors relied on this ability to eat and to avoid being eaten. When they needed to spot well-camouflaged birds and rabbits hiding in the bushes in order to avoid starvation, this ability to see things through clutter was crucial. It was no less important, of course, for them to identify the vague outline of a predator hiding in ambush in order to avoid becoming dinner in the short term and avoid extinction in the long term.

Although city dwellers rarely, if ever, find themselves in need of locating a predator lurking in the distance, this type of seeing still seems to be deeply ingrained. For example, a friend of mine who does long-distance open-water swims in the Caribbean told me that she does not suffer from shark paranoia, but, regardless of how she feels about it, her brain doesn't take sharks lightly. When her head is under water during long swims, she says her brain's visual system seems to be scanning constantly for the outline of sharks in the distance. And it often "sees" shark outlines if something even comes close to the shark template, she explained, although 95 percent of the time it's nothing, not even a fish.

Michael Shermer, founding publisher of the excellent magazine *Skeptic*, has studied and written about this skill/habit/obsession of our brain for many years. He calls it *patternicity*. Defined as "the tendency to find patterns in meaningless noise," Shermer says our brains do this so often and so well that they ought to be thought of as "pattern-recognition machines."[3] It's all good, of course, right up until patternicity starts pushing us beyond reality and causing us to see too many things that do not exist. That's when we get ourselves into trouble.

If I were walking in a forest at sunset and there was by chance a shadow that was vaguely shaped like a bear, my brain might instantly serve up a more complete and convincing image of a dangerous bear lurking in the dark. He's not there, but I just saw him. *I swear, I could even see his sharp teeth and menacing eyes!* Now, if a bear really was there, my brain might have saved my life by alerting me. If no bear had been there, however, I was startled briefly and it's no big deal. But what if I was certain I had seen Bigfoot, a demon, an alien, or a god? That might complicate my life unnecessarily. Shermer explains:

> Unfortunately, we did not evolve a Baloney Detection Network in the brain to distinguish between true and false patterns. We have no error-detection governor to modulate the pattern-recognition engine. Thus, the need for science with its self-correcting mechanisms of replication and peer review. But such erroneous cognition is not likely to remove us from the gene pool and would therefore not have been selected against by evolution.[4]

On one hand, it makes sense for us to see some patterns of things that aren't really there in order to be very good at seeing real ones that matter. On the other hand, we need to be aware of this phenomenon because it can lead to a confident belief in things that are not real or true. Additionally, patternicity is not limited to vision. It impacts hearing and thinking, too. Good skeptics understand how the brain often creates false patterns, so we know to be very cautious when considering claims of UFO sightings, for example, or anything else that is unusual. It only makes sense to be skeptical and ask for additional evidence when people claim to have seen or heard extraordinary things. Maybe they did, maybe they didn't. Given what we now know about the brain, however, are you going to believe someone who tells you she saw a flying saucer or Bigfoot last week? She doesn't have to be lying to be wrong. Anyone with perfect vision can see poorly. Anyone with a bright brain can think and come to the wrong conclusions. Anyone with an excellent memory can have wildly inaccurate memories.

THE APE THAT WASN'T THERE

As if all this weren't enough, we also have to be on guard against something called *inattentional blindness*. This fancy phrase is what scientists call it when people look right at something and think they are paying close attention, but really aren't seeing as well as they imagine. Inattentional blindness is the reason it's not safe to drive while talking on a phone. You can be staring wide-eyed at the road ahead of you, but as your brain devotes itself to the phone conversation, you fail to notice the truck that brakes in front of you or the motorcycle rider who pulls out in front of you.

Scientists Christopher Chabris and Daniel Simons constructed a funny and effective experiment that shows just how easily inattentional blindness cripples our awareness. If you would like to take the test yourself, visit their website and view the brief video before reading the rest of this paragraph (www.theinvisiblegorilla.com/videos.html). What Simons and Chabris have revealed about us would be difficult to believe if they hadn't repeated the experiment so many times with similar results. Their video shows two teams of people dribbling and passing basketballs. One team is dressed in white shirts, the other in black. Test subjects viewing the video are instructed to count the number of times players in white shirts pass the ball as they move around on the court. That's it, nothing more. But while test subjects are concentrating and trying their best to keep an accurate count of the passes, a woman in a full-body gorilla suit walks into the middle of the scene, beats her chest, and then casually walks off. To be clear, it's not some brief, subliminal flash of an image. The gorilla is on screen for a total of about nine seconds. After researchers ask for pass counts, they ask test subjects if they happened to have noticed anything unusual or out of place during the video. Surprisingly, about half of the people who take this test *don't see the gorilla*. They fail to notice a hairy gorilla strolling around right in front of their eyes for nine seconds! They are so focused on the ball that their brains fail to recognize and remember the gorilla, even though the ball they are intently tracking with their eyes passes directly in front of the gorilla.

This is inattentional blindness, and it's a normal, useful feature of our brains. We can't pay attention to everything all the time, so our brains zoom in on some things while sacrificing broader awareness. The implications ought to be obvious to aspiring skeptics. Not understanding inattentional blindness can lead us to be overconfident in our ability to take in everything around us and not let important things slip by. The truth is, however, we do miss important and unusual details— such as gorillas!

We can't always be sure that we see everything we need to see in order to make sense of a scene or a situation before us. If someone observes what she thinks is a ghost, for example, the "ghost sighting" might hinge on key details in the scene being left out by her hyper-focused brain. For example, maybe the brain was so locked in on what she thought was a ghost that she failed to notice the lawn lights a few feet away that were shining on the bushes at an angle that formed a light image similar to what popular culture had taught her a ghost is supposed to look like. Inattentional blindness is yet another reason to be skeptical of eyewitness accounts and to always ask for better evidence.

CONFIRMATION BIAS

Anyone who hopes to be a good skeptic has to understand "confirmation bias" and consciously resist it. You will never defeat it totally, but knowing it's there can prevent you from becoming a slave to it and turning your head into a haven for false beliefs. Confirmation bias is the problem we all have—even good skeptics—when it comes to thinking about our beliefs. Without being aware of it, we tend to protect our beliefs. Confirmation bias draws us toward evidence and arguments that support our beliefs while simultaneously turning us off to evidence and arguments that go against our beliefs. It might seem obvious that people gravitate toward the things they already agree with and like the arguments they already agree with, but it's not obvious when it's happening to you. This bias makes believing in weird, unproven claims much easier than it would be otherwise. It can make recognizing that a belief is wrong and abandoning it extremely difficult. Someone who believes in psychics, for example, has a brain that will naturally perk up and go into overdrive when someone comes on TV to talk about "the confirmed science of psychic readings." The stories and "facts" presented by the person on TV might be marked "high priority" by the brain and stored for future retrieval. But if a skeptic like me comes on TV and begins explaining how psychic readings really only *seem* to work and why it doesn't make sense to believe in them, the believer's brain is likely to instinctively go into siege mode. The drawbridge is raised, crocodiles are released into the moat, and defenders man the walls. Little or nothing the skeptic on TV says gets in. Anything that does somehow manage to penetrate is likely to be promptly forgotten forever. The worst part of all this is that the believer usually doesn't recognize how biased and close-minded he is being. He likely feels that he is completely rational and fair. It doesn't happen just with fans of psychics. We are all vulnerable to this distorted way of thinking.

A good skeptic seeks to keep an open mind and knows that it's always wise to second-guess her conclusions and beliefs, no matter how sensible they seem or how beloved they may be. We have to do this both because we aren't perfect in our thinking and because things change. If you think you will never make a single mistake in your life, then, by all means, never reconsider your conclusions. But if you are like the rest of us, then you want to constantly check yourself. Remember, nothing is static. New information and new discoveries often mean some things were in error and need revision or rejection. Nothing should be thought of as written in stone. Sure, some conclusions are very likely to be true for all time—but which ones? We can't possibly know for sure in every case. The best we can do is accept conclusions that are backed up by the best evidence we have today and agree to change our minds if better evidence ever comes along that says something different tomorrow. You never want to be too loyal to a claim, a belief, or an conclusion about something. I think the claims of astrology are untrue, but I'm not chained to that position until my final breath. If the scientific process ever shows that Libras are the best lovers and Capricorns make better gardeners because of star and planet locations, then I'll adjust my view of astrology immediately. Changing your mind is okay. It shows wisdom and maturity. Never wavering from a belief no matter what is a mistake. What matters is trying to be as closely aligned with truth and reality as you can be at all times. And that requires many navigational adjustments over a lifetime.

THE TALE OF LITTLE GRETCHEN GREENGUMS

The following short story is presented to help you remember some of the weird ways our brains can cause us to believe in things that aren't real.

It's about a good girl who was not a good skeptic and suffered a most grisly fate as a result: brain rot and a derailed life. Please remember this tragic tale whenever you find yourself on the brink of accepting an extraordinary claim that lacks overwhelming evidence. Maybe it's true, but do make the effort to think before you believe. Recall what happened to little Gretchen Greengums whenever you are tempted to believe in something unusual for no good reason.

NO DOUBT about it, young Gretchen Greengums was a winner. Hands down, she was the smartest and most popular kid in the tiny town of Rottbridge, population 7,495. Greengums was special, and everybody knew it. Cute as a button, too. Nobody cared that she had a wicked overbite and a frightening set of eyebrow ridges that might have looked a tad thick on a Neanderthal boxer. There was just something special about the girl, and the whole town loved her for it. Gretchen was so beloved and respected that no one bothered to oppose her when she ran for president of the fifth grade class. She was the rare overachiever no one resented. Many adults predicted that she was sharp enough and motivated enough to make it all the way to the top. More than a few openly talked about her having the potential to one day run the town's most prestigious and successful business, a factory that recycles used kitty litter to make dentures and decorative tobacco pipes. But Gretchen had even bigger plans than that.

Her goal was to head off to college after high school and earn a degree in astronomy. She wasn't shy about telling anyone that she was going to either prove or disprove string theory *and* crack the mysteries of both dark matter and dark energy once and for all. Nobody in town had the slightest clue what she was talking about, but they knew enough to respect her ambition and not to doubt her.

GRETCHEN WAS NO SPAZ. In addition to her lofty academic goals, she also dreamed of being the best baton twirler in the world. She already had won the annual Pro-Am Rottbridge Baton-Twirling Championship three years straight. Everyone just knew she would go pro one day. All she needed was a little more seasoning, a bit more height on her back throws, and she would be ready for the big leagues. Yep, she was the real deal, the total package, the golden child of Rottbridge. Gretchen Greengums was going places. But, out of the blue, everything changed one strange autumn day.

During a twirling practice session all by herself out on the YWCA field, Gretchen tossed her baton high in the air like she had done a million times before. But this time, while looking up, something odd caught her attention. Up in the sky, something was hovering in the air. It was strange. *Definitely no plane*, she thought. As the baton fell to earth beside her, she just kept staring up. "Cripes! What in heckfire is that thing?"

Alternating jolts of fear, curiosity, and excitement surged through her little body as she began to figure out what it was. "Wait, I know! It's a UFO! Just like the one I saw in that show on the Pseudohistory Channel last week! Holy moley! It's even shaped kinda like one of the spaceships they showed! It must be aliens! What else it could be?"

> *Believing is seeing. We don't really see with our eyes as most people assume we do. We see with our brains because our brains construct a scene based on images that come in through the eyes. It's never a perfect reflection of what's actually there in front of us. Furthermore, what the brain sees can be strongly influenced by what the brain already believes. Even if she wasn't completely convinced by it, Gretchen's recent viewing of a nonscientific TV show about UFOs may have primed her to "see" something that wasn't there.*

Gretchen analyzed the object carefully. Her mind raced. She placed her hands on top of her head and said out loud that this was the most important day of her life. No longer hovering, the object began moving away toward the distant horizon. "It's gigantic," she said aloud to herself, "Maybe 1,000 feet wide! No way is this a balloon or a blimp. It's gotta be a spaceship. Gotta be!" She began to imagine what the occupants might look like. Questions flooded her mind: *What is their planet like? How smart are they? Are they friendly? Will they share their knowledge with us? What if they're mean? What if they wanna eat us or something like that?*

Our brains find meaningful patterns, even when none are there. When we look around at the world, our brains naturally search for patterns and try to connect the dots in visual noise or clutter to form whole images—even when the dots don't really connect and there is nothing actually there. In this case, Gretchen is looking at nothing more than a flock of birds in the distance, a likely cause of many UFO sightings. Without her being aware of it, Gretchen's brain instantly connected the dots and created one solid image out of several individual birds flying in formation. Furthermore, the late-day sun is reflecting off of the birds' bellies, making them appear unnaturally bright. She thinks it is so big because it's difficult, if not impossible, to accurately estimate the size of something in the sky if you have no idea what that something is.

GRETCHEN WAS NO IDIOT. She remembered something her third-grade teacher once said to the class about being skeptical of UFO sightings. She blinked hard and slow twice, took a deep breath, and tried to look at the scene before her with fresh eyes. She clenched her teeth in total concentration and stared at the UFO with all the mental focus and intensity she could muster.

But it was still there. "No!" she said, "This really is real. I'm not imagining it and there is no way that big thing is a plane or a helicopter or anything else like that! I can't figure out what it is. It has to be an alien spaceship!"

We don't see everything we look at. We can stare at a scene and believe that we are paying close attention to it but still miss important things going on right in front of our eyes. Inattentional blindness is a standard problem that comes with human brains. Remember the gorilla experiment cited earlier in this chapter. Gretchen is so focused on the "spaceship" that she fails to notice another flock of birds close by that fly by right in front of her field of vision. If she had noticed them, she likely would have realized what it was she was seeing in the distance. Notice that Gretchen also makes the mistake of thinking that her being unable to identify the object is somehow a good reason to conclude it must be an alien spaceship. A good skeptic would have opted for "I don't know."

Gretchen continued to watch the object for another few minutes. But as the Sun set behind her, the UFO vanished. Stunned by what she had just experienced, she took a deep breath. "Whoa! I just saw a UFO, a spaceship from outer space!" Her hands trembled, but she wasn't scared anymore. Now she was excited. *This is fantastic*, she thought. *Now I know for sure. Aliens are real! We are not alone!*

Gretchen took off for home, running as fast as she could.

Smart and honest people are sincerely wrong all the time. Most people who claim to have had a paranormal or supernatural experience are probably not lying or suffering from a serious mental illness that might have caused them to experience what they did. Most likely, they really did see or experience something. But that "something" is probably an internal brain experience of one kind or another that normal people can and do have when their brains misinterpret real events or confuse fantasy for reality. For this reasons, it is not only the claims of others that we need to be skeptical of. We also must be skeptical of our own thoughts and perceptions.

When she arrived at home, Gretchen told her family about the UFO. The next day, she told everybody at school about it. Because of Gretchen's reputation for being smart and honest, everyone was inclined to believe her. "Wow," said her teacher, Mrs. Flurston. "I guess this means alien are real. How exciting! Hey, students, what do you say we do a special lesson on UFOs this week?"

"Yay!" yelled the students. Mrs. Flurston then told the class that she would bring in a few "excellent and educational" Pseudohistory Channel DVDs she had at home. She said one shows how Hitler had used alien spaceships to nearly win World War II. The kids cheered again, even louder this time. "Another one of the DVDs explains how aliens visited us thousands of years ago and left a few pets behind by mistake," said Mrs. Flurston. "Those pets, it turns out, were our distant ancestors!" This time the students all stood up and clapped wildly.

Little Kevin Handfurt, the cute class clown with a weak bladder, raised his hand enthusiastically and blurted out that he could bring in his daddy's "science books" about the UFO crash at Roswell. "It's even got real pictures of the alien bodies!" he said. Mrs. Flurston nodded approvingly. The boys in the front row high-fived each other. Some of the kids punched the sky in celebration. Even Todd McSfinster, the grumpiest kid in class, got in the spirit. "Finally!" he said, "We get to learn some real science. It's about time!"

One classroom couldn't contain this much energy for long. By the end of the week, the whole school was buzzing about Gretchen's UFO, and soon after that, the whole town was. When Gretchen got home from school one day, she was greeted in the living room by a reporter from the local newspaper who had been waiting to interview her. The very next day, an article about her was on the front page of the town paper. It included a sketch of the spaceship, based on Gretchen's description. The illustration included several details that Gretchen didn't actually say that she saw, but the artist thought it would make it more interesting for people so he added them. She never saw any strange markings on the side, missile things attached to it, or an alien waving from a window like the newspaper sketch showed, but the general shape of the object was about right, so Gretchen didn't say anything.

The last sentence of the article read as follows: "The town of Rottbridge can only applaud Gretchen Greengums's extraordinary courage for sharing this story of her close encounter. While there are many questions still left to be answered, we now know, thanks to this one special little girl, that we are not alone in the universe."

Even better than the article, Rottbridge's mayor, the Honorable Earl Wayne Chipply, publicly and officially praised Gretchen at a special town-hall ceremony. He even presented her with a fancy plaque suitable for hanging. During the event, the mayor called her a hero. He also said that he too was a believer and encouraged everyone to keep watching the skies because "they are definitely up there."

Reputation and titles aren't everything. When it comes to extraordinary claims, a person's credentials and reputation matter, but only so far. These things aren't nearly as important as good evidence. Always remind yourself that the person making an unusual claim—no matter how impressive his or her accomplishments or reputation for honesty may be—is human and therefore is capable of mixing up fantasy and reality just like anyone else. Also be on guard about placing too much trust in the news media. Journalists are human, too, and that means beliefs, biases, and weak skepticism can and do creep into their work. We also must be careful about the natural human tendency to look up to, follow, and believe authority figures. Just because people may be in positions of power and trust, such as a mayor or an educator, doesn't mean they can't fall for bogus beliefs just like anyone else.

TIME FLIES FOR HEROES. Fifteen years after seeing the UFO, Gretchen was still smart and still cute. But she was not quite as popular as she used to be. Things were never the same for her after that day. She became convinced that the spaceship would come back, so she spent more and more of her time looking up and waiting. Over the years, the townspeople, then her friends, and finally her own family grew bored with the story. They still liked her well enough, and most people thought Gretchen meant well, but after years of UFO hunting with nothing to show for it and all the constant talking about aliens, the whole thing grew stale and a bit embarrassing. There was just something about a young woman in the prime of her life, with so much potential, constantly staring up at the sky and doing nothing else that some people didn't feel comfortable with. But she didn't care. Day after day, night after night, Gretchen stood with her head facing up, scanning the sky, looking for that UFO.

"Look down, once in while!" her father screamed at her one day. "For Pete's sake, you're wasting your life staring at clouds all day!" But Gretchen kept looking up and hoping the aliens would return. Sadly, all that looking up made her lose sight of her dreams down here on Earth.

Gretchen never did go to college. Never took over as boss at the kitty-litter recycling factory. As soon as she graduated from high school, she devoted herself full-time to "alienology." She devoured every book, magazine, and website she could find that promoted belief in UFOs and aliens. She read about alien abductions, the famous crash at Roswell, secret activities at Area 51, and people who were in communication with aliens. She attended lectures by people who said they had been aboard UFOs and taken trips in space. She partied with Raelians one wild weekend in Vegas. She played her *Close Encounters of the Third Kind* DVD so many times that one day her DVD player just up and died.

Gretchen's greatest moment of glory probably came when she appeared for twenty-seven seconds in a primetime Pseudohistory Channel documentary titled *Nostradamus: Human Savant or Alien Astrologer?* She told them her UFO story on camera, but it didn't make the final cut. They only used her comments about aliens probably having the ability to predict the future and her statement that therefore there's a good chance that Nostradamus was one of them or at least consulted with them.

Gretchen had no interest in reading books that were skeptical of UFO claims and refused to listen to the perspectives of professional astronomers, Search for Extraterrestrial Intelligence (SETI) researchers, astrobiologists, and National Aeronautics and Space Administration (NASA) officials on this subject. She felt she knew better than all of them. After all, *she* had seen a spaceship with her own eyes—not them. To her, it made no sense to waste time considering possible alternate explanations, so she devoted herself to researching stories and mining data only from those who believed. Her older brother was skeptical and often tried to argue with her about UFOs, but she wouldn't have it. "He didn't see, so he doesn't know," she said to herself. "Besides, it's not like he can disprove my story."

> *Confirmation bias can prevent us from thinking clearly about our beliefs. We all tend to pay attention to and remember information that supports our beliefs while shying away from and forgetting information that challenges or contradicts them. Everyone does this, which is another reason why being a good skeptic is so important in everyday life. If you care about truth and reality, if you don't want to waste your time believing nonsense, then you must be aware of confirmation bias and resist it. You have to work at keeping an open mind and considering other ideas and evidence, no matter how foreign or uncomfortable they may feel. It's also not the skeptic's job to disprove extraordinary claims. It's the responsibility of the person who makes the claim to prove it.*

That big UFO day was a long, long time ago. Gretchen's gone gray now. She is winding down a life that certainly no one would have predicted for her half a century ago. She never did leave Rottbridge. Never did take a stab at string theory. Despite all that potential, she never competed in another baton-twirling event after fifth grade. To this day, she still walks around town with her face angled upward. When she talks to people on the street, she rarely makes eye contact. Too busy looking up. It wasn't only her social life and career that suffered. Gretchen paid a physical price as well. Today her legs are gnarly sticks of scarred flesh. All that staring up at the sky doesn't make for safe walking down on the ground. Over the years, she's slammed into so many fire hydrants and park benches and collided with so many skateboarders that it's a wonder she's still alive. If you didn't know better, after one look at those mangled and dented shins, you might assume she's a retired Muay Thai fighter who had spent a long career battling in Bangkok. Yeah, that day Gretchen saw her UFO changed everything about her.

Gretchen spends most of her time these days alone in a small apartment, writing and posting her UFO newsletter on the web. She'll play that DVD of her twenty-seven-second Pseudohistory Channel sound bite for anyone willing to watch. A couple of times a year, she gives a lecture on "UFO Science" at the Rottbridge public library. Pretty much the same small audience shows up every time. There are the diehard believers, the homeless people, and a few who get lost looking for the romance-novel section. But Gretchen never fails to deliver. When it's show time, she glows with enthusiasm. "I *know* intelligent extraterrestrials are visiting us," she says. "I saw them with my own eyes. They are here."

HER UNBLINKING STARE IS CREEPY TO SOME. But others seem inspired to believe. "I can remember that day so clearly, all those years ago," she says. "The spaceship hovered right above my head. It was silver with flashing lights. It had strange markings on the hull. There were missiles or some kind of auxiliary rockets on the sides, too. I could even see a window, and I could see an alien with a large head inside. He was very pale, either gray or white. I'll never forget his eyes, large, black eyes. And then he waved at me. Yes, it was a moment I'll never forget."

> *Science has shown that our memories are unreliable. Our brains do not provide accurate replays of past events for us to watch. They construct our memories from bits of information. When doing this, our brains edit and embellish. Things are left out, things are added in. Sometimes other memories are mixed in. Sometimes things we saw in a movie, watched on TV, or read in a book are added to the mix. And all of this can happen in a way that feels right to us. Memories of things that never happened can feel 100 percent accurate. This is why confidence, honesty, intelligence, and reputation have little to do with whether or not a person is remembering something accurately. Trust people, but don't trust their memories.*

Gretchen checked out at the age of eighty-seven. Right up to her death, she clung to her belief that she saw a spaceship long ago. It never occurred to her that she could have reacted to that event differently. She never imagined that her UFO was nothing more than a flock of birds, the setting Sun, and her own brain conspiring to fool her. She had never heard of confirmation bias or inattentional bias. She never learned that memories are fallible fabrications of the brain. No one ever advised her that it is better to live with an unanswered question than to pretend to know something you don't. No one had ever suggested to that bright little girl that it was important for her to think before she believed.

The loss goes two ways. It wasn't only Gretchen who paid a price for not being skeptical enough. Everyone else lost too. Gretchen's family lost the joy and reward of watching a daughter strive for big things. The town was robbed of one of its brightest lights. In real life, it works this way, too. Our entire world is diminished when minds, filled with potential, are sacrificed on the many altars of irrational belief. Sadly, it happens every minute of every day. Millions of people who are much like the fictional Gretchen have their wonderful and powerful brains smothered and dimmed as their skulls are packed with the cheap sawdust of unproven and unlikely claims. Given the many very serious problems humankind currently faces and will face, I'm not sure that we can afford to keep losing so much intellectual wealth each generation. It seems to me that we need all the brainpower we can muster, including yours.

YOUR BIZARRE AND BIASED BRAIN

The following is a list of some natural mental biases and common thinking errors that can lead any one of us to accept wrong or dubious claims and beliefs. To become a good skeptic, you don't need to earn a degree in psychology, but you should have some level of awareness about how your own thinking can trick you. You can never be immune, but the more you know, the safer you will be from self-deception. It can also help you to understand how others may have gone wrong in their thinking when they are trying to convince you of something they believe. Billions of smart, sane, and sincere people fall for nonsense every day because of these weird thought processes.

It may startle some to realize it, but a very small portion of our brain's activity is what you might describe as conscious, or things we are aware of. Probably more than 90 percent of what is going on in our brains every day is *subconscious*, meaning we are mostly clueless about it. Think of it like this: Our brain is like a house that has twenty rooms on four floors. But we are allowed to enter only one room in the entire house. The doors to the other nineteen rooms are locked. The house is always busy. A lot of activity is going on in those other nineteen rooms and it all directly impacts us. But we don't get to peek in.

Basic brain-awareness information should be taught to everyone and taught early. Sadly, however, the vast majority of the world's people go their entire lives without ever learning about it. This is one of the reasons why so many brains end up in all the wrong places. But not yours, right?

- **Anchoring.** This bias leads us to rely so much on one past experience or one piece of information that we ignore or reject new, more, and better information that contradicts it. Don't drop anchor in the muck of a past error when you could set sail for more sensible lands.
- **Argument from authority.** We are social creatures who devote a lot of our time to thinking about social ranking. A by-product of this obsession is a tendency to blindly trust the claims and ideas of people who rank above us. Whether or not we realize it, perceptions of authority and superiority impress us and influence us whether they are valid or not. A good skeptic knows to assess the merits of a claim independent of who presents it. Just remember that truth can come from below us and lies can come from above us. Pay less attention to the messenger and more attention to the message.
- **Argument from ignorance.** When you don't know, you don't know! Too often, many people think that the absence of a normal, sensible answer must therefore mean that the answer is something special, magical, or supernatural. No, unknown means *unknown*. When weak skeptics say, "You can't explain exactly how it happened, so it must me a miracle," a fair reply is: "You can't explain exactly how the miracle happened, so it must be a natural event."
- **Availability cascade.** When you say something repeatedly, you are likely to believe it with increasing confidence, *even if it's not true*. It turns out that we actually listen to ourselves when we

speak—and trust what we hear. So be careful about what you go around saying. Make sure it's valid, because your brain is always listening.

- **Availability heuristic.** This is a big one. We tend to be influenced far more by one or two "real-life" examples in our head than we are by more abstract facts and statistics. So if you were to show a person a few studies done by credible researchers that showed ear candling failed to deliver health benefits, a weak skeptic might not be impressed because she's too busy thinking about a family member who swears ear candling cleared his sinuses and fixed his vertigo problem.

- **Backfire effect.** I observe this one in action all the time, and it drives me nuts. It turns out that there is a natural bias that can make us believe something with greater confidence when we are confronted with evidence or strong arguments that oppose it. Because of this bias toward digging in your heels, more and stronger evidence often results in strengthened irrational belief. Be aware of it and resist it or you may end up defending nonsense.

- **Base-rate fallacy.** This one derails us with ease. We can readily find ourselves focusing on one tiny speck of information (a single story, for example) or on bad data that supports a claim while simultaneously ignoring more credible information or a larger body of data that goes against it.

- **Bias blind spot.** This bias can trip anyone, of course, but I always stress to skeptics that they pay particular attention to it. We can more easily see biases and illogical reasoning in others than we can in ourselves. It seems natural to think you are less likely to be fooled than everyone else around you is. That is the kind of arrogance that will lead you straight into the spider's web.

- **Confabulation.** This is the scary weakness we all have for remembering things incorrectly yet feeling like the memory is totally accurate. As I've described earlier, it's very easy for the brain to jumble the timeline of events or combine different events to create one believable memory. Good skeptics have to keep this one in mind because very intelligent, confident, respectable, and sincere people can be completely wrong when they describe something that they experienced in the past.

- **Confirmation bias.** As explained previously, this wicked mental filtering process will have you noticing, emphasizing, and remembering information and experiences that support your beliefs while ignoring and tossing out data that contradict your beliefs. Make the effort to pay attention to this tendency and always consider evidence and arguments that go against the grain of your brain. A good skeptic keeps an open mind.

- **Conformity, also known as the bandwagon effect.** You probably think you are an individual who thinks independently and doesn't care too much about what everybody else thinks or does. The truth is, however, we all feel the lure and often succumb to social pressures when it comes to how we eat, dress, talk, buy things, entertain ourselves, *and think*. A good skeptic remembers this when doing an inventory of his or her beliefs and conclusions. *Did I really think this through, or am I just following the herd?* Nothing to be ashamed of here. We all want and need the herd to some degree. Just don't allow yourself to be blinded by this powerful and constant pull on your brain.
- **Emotional bias.** This one is so simple but can cause many problems. Our emotions can easily dominate our attention to the point where we don't see or consider relevant information right in front of our faces. Emotions are great, the spice of life, but don't let them rule your decision-making process. Don't just feel, think!
- **False-consensus effect.** This is the tendency we have to overestimate how many people agree with us about the things we believe. This can cause people to have more confidence in their beliefs. *If everyone believes in ghosts, then they must be real.*
- **False memory.** This is a common problem in which imagination is mistaken for the memory of something that actually happened. This one has obvious implications for everything from angel sightings and near-death experiences to alien abductions.
- **Forer effect, also known as the Barnum effect.** This bias helps explain the popularity of things like psychic readings and astrology. When given a general personality description, for example, many people naturally think it describes them in a unique and special way, even though it is vague and could apply to many people or even to all people.
- **Framing effect.** This bias scares me as an author. We can be swayed to judge information and ideas based on who delivers them to us and how they are presented. So, if you don't like me, you may not be inclined to accept my message that skepticism

is good for you. I hope you like me. But even if you do like me, don't let that be the reason you embrace skeptical thinking. Do it because you thought it think it through and it makes sense.

- **Hallucinations.** It is well established that human beings are capable of seeing things that aren't there. What is really weird is that hallucinations can occur in the same parts of the brain that see and hear real things, so hallucinations can feel totally authentic. The obvious question is, why do so many people who go on and on about paranormal/supernatural sightings fail to consider hallucinations first? If someone says, "It was too real to be a hallucination," she or he has done nothing more than reveal that she or he doesn't understand hallucinations.
- **Hindsight bias.** Never underestimate how effective our brains can be at keeping us feeling good about ourselves. This bias is the routine lie we tell ourselves and then believe when we learn that we were dead wrong about something. "I knew it all along," is an ego-soothing, natural reaction. But you probably already knew that, right?
- **Illusion-of-truth effect.** When we hear a claim that is familiar to us, we are more likely to believe it than we would if we had never heard it before. This means simply being exposed to state-ments about "X" makes you prone to trust and believe in "X" further down the line when told about it. It can work with ESP and conspiracy theories the same way it works with car commer-cials and laundry-detergent ads.
- **Illusory correlation.** This is very common. Remember how we love to connect those dots? Well, our brains are so eager to do this that many of us often connect two or more unrelated events and then wrongly conclude that one caused the others or that they are somehow linked in a meaningful way when they are not.
- **Inattentional blindness.** We are capable of staring at some-thing, thinking that we are paying very close attention to the scene, but still missing important or unusual things that are there right in front of our eyes. Remember the gorilla that pounded its chest in the middle of a basketball game.
- **In-group bias.** Everybody knows about this one, but how many make an effort to recognize and resist it when confronted with an extraordinary claim? We tend to respect, admire, and listen to people who are in our groups (nations, religions, schools, clubs,

etc.). This can lead us to trust them about a belief or claim over someone who is outside the group, even if the outsider makes more sense.

- **Irrational escalation.** This one can be uncomfortable to deal with, but fight it if the situation demands. This bias can leave us feeling too invested in a decision or a belief to let go and reverse course, even when logic and evidence make it clear that we should. Please don't spend years or an entire lifetime saddled with a bad belief. If you figure out that it's wrong, cut your losses and move on.

- **Observational-selection bias.** I notice this one going on in my own head all the time. Whenever I buy a new laptop, MP3 player, or some other gadget, I suddenly see that specific model everywhere. Of course, it's because I have a heightened awareness of it because I just purchased one. This selective perception can fool us into thinking special or weird things are happening when they aren't. *Oh my gosh, so many people at school today are wearing a crucifix necklace just like mine. It must be a sign.*

- **Pareidolia.** This one helps to explain many claims people make about seeing strange things. When we seė or hear something that is nothing meaningful, our brains might decide to go ahead and give it meaning anyway. Our brains can create a recognizable picture in a cloud or on a slice of toast. We can also hear words in meaningless noise where none exist.

- **Priming.** Never underestimate how easy it is to manipulate a human being. For example, just by hearing or reading a word, you are likely to think of it and to be influenced by it later in unrelated thoughts and activities—even if you forget all about originally hearing or reading that word.

- **Status-quo bias.** We are creatures of habit. I don't know about you, but I haven't changed my hairstyle since first grade. A general reluctance to change can derail us when we try to sort out truth from fiction. If we have a few bad beliefs in our heads, we may hold onto them for no other reason than we like familiarity and don't like change.

Got all those memorized? I hope you remember them, because there will be a test—it's called the rest of your life. No, you don't need to memorize them. But it helps to familiarize yourself with them, and you should at least be vaguely aware of what your subconscious brain is up to every day. Accept that we need these strange biases and weird mental processes. If not for the kooky minefield of tricks, shortcuts, and deceptions inside our heads, we would not function as well as we do in our daily lives. It's an interesting twist: If our brains were not so nutty, we'd all go nuts. In most cases, the mysterious workings of the brain are necessary for efficient thinking and living. So don't hate your brain for its deceptive ways in setting you up to make mistakes in reasoning and perception. Thank it for doing the best it can to get you through one more day safely. Just be sure to understand that you can never trust it completely when trying to separate truth from fiction.

ARM YOURSELF IN THE WAR AGAINST REALITY

Now that we know something about why it's important to be skeptical and we looked at some of the ways our brains can trick us into believing unlikely things, let's turn our attention to some specific challenges in the next chapter. It helps if a good skeptic is familiar with key problems associated with some of the more popular beliefs out there. For example, you would be right to ask, "Where's the evidence?" if someone tried to convince you that the lost continent of Atlantis was real and was inhabited by an advanced race of superhumans or extraterrestrials. But it would be much better if you had enough background information to be able to ask the Atlantis believer a few good questions that both expose the weakness of the claim and encourage the believer to think more deeply about it. Sure, it's correct that no one has proved the story of an alien spaceship crashing near Roswell, New Mexico, in 1947. But it's much better if you can share the real story of what happened there. When someone attempts to get you to buy into the latest magical end-of-the-world prediction, you can deflate his sales pitch faster and more effectively if you know something about the long history of these baseless beliefs. The next chapter is going to be a wild ride around, over, and through tall tales and wild claims, so strap on your helmet and brace yourself. We're going in.

GOOD THINKING!

- Don't forget this about your memory: You can't trust it! Human memory is a fragile and fallible thing. Our memories are *constructed* in our head, like stories based on bits of information. They are not reliable playbacks of "recordings."

- Your brain *constructs* and *interprets* what you look at. You don't have a camera in your head that faithfully captures reality. What we see is something the brain has produced for us, based loosely on what comes in through our eyes. It's never 100 percent true and complete. For this reason, we can't always be sure about what we think we see.

- Learn about some of the many cognitive biases that come with every human brain. If you are not careful, these weird thinking processes will trip you up, leading you to make errors in reasoning and embrace bad beliefs.

Chapter 3

A THINKER'S GUIDE TO UNUSUAL
CLAIMS AND WEIRD BELIEFS

Now that we have an idea of just how weird and troublesome our own brains can be when it comes to sorting out truth from fiction, let's consider some specific claims that are unproven and unlikely to be true even though millions of people confidently claim that they are. As we explore these claims and expose them to the light of skepticism, please keep in mind that in most cases a good skeptic does not claim to *know* that an unusual claim is not real or true. A very important part of being a good skeptic is maintaining an open mind.

It goes both ways—just as believers shouldn't pretend to know things that they actually don't, skeptics ought to avoid following this kind of thinking, too. For example, based on the absence of good evidence and the improbability of a large creature eluding scientific confirmation in a confined space for so many years, I'm very confident that the Loch Ness monster doesn't exist. But I stop just short of declaring that it's impossible. After all, stranger things than that have happened. Microscopic mites, germs, and walking on the Moon once seemed pretty crazy, too. The Loch Ness–monster claim doesn't necessarily contradict the known laws of nature. And I'm not emotionally invested in a world with no Loch Ness monster. I wouldn't shed one tear or feel awkward for one second if Nessie were netted tomorrow. In fact, I would be thrilled to hear the news. Sure, it would mean that I had been leaning the wrong way on this one, but so what? I don't care if I was wrong yesterday; I want to be right today. The discovery of a modern-day plesiosaur would be wonderful, something to celebrate. In fact, I would probably be on my way to Scotland within hours of the announcement so that I could try to interview the discoverers. A good skeptic does not stubbornly defend a rigid position against a claim at all costs. A good skeptic only wants to base her or his conclusions on the best available evidence and arguments. If truth and reality were to translate to a world with ghosts, magic crystals, and vampires in it, then that is the world good skeptics want to know about. Skeptics are often accused of being "against everything" when in reality we are against nothing except mistakes, delusions, and lies. Why should anyone disagree with this position? Furthermore, shouldn't we all be against mistakes, delusions, and lies?

Belief on the Brain
in the United States

A UFO crashed at Roswell	21 percent	65,921,948 people
Vaccines are linked to autism	20 percent	62,782,808 people
Obama is the Antichrist	13 percent	40,808,825 people
Bigfoot exists	14 percent	43,947,966 people
Earth and human species are less than 10,000 years old	46 percent	144,400,458 people
The Moon landings were faked	7 percent	21,973,983 people
Ghosts	42 percent	131,843,896 people
Astrology	26 percent	81,617,650 people
Atlantis	41 percent	128,704,756 people
Miracles	76 percent	238,574,670 people
UFOs	40 percent	125,565,616 people
Reincarnation	20 percent	62,782,808 people
ESP	41 percent	128,704,756 people

Sources: Gallup, Public Policy Polling (2013), Baylor Religion Survey (2006), Harris Poll (2009). Illustration by Kevin Hand.

Too many good skeptics fall short when talking with believers about their beliefs. It is not enough to simply demand evidence and poke a few holes in a claim. Experience has taught me that a good skeptic who wants to be helpful and make a lasting impression on others needs to bring something more to the party. Relevant ideas, a bit of historical context, some science, some insight into how the brain works, and maybe a good alternate explanation can go much further than ranting about the absence of evidence and leaving it at that. If being constructive and getting people to think more seriously about their beliefs is the goal, then arm yourself with the following skeptical summaries of popular beliefs. If they make sense to you, please share them with anyone and everyone willing to listen. If you happen to believe in some or all of the following claims, I hope you will keep an open mind as you read and give my words fair consideration. No less important, I hope you receive them in the spirit they are offered. I'm not trying to convince you of *the truth according to Guy*. I'm attempting to help you think for yourself in more effective ways and then draw your own conclusions. So keep reading and keep thinking!

MAGICAL, SUPERNATURAL, PARANORMAL

One of the most consistent things about people everywhere is *belief in unusual things that probably are not real or true*. It may not be universal, but it's close. It's as if we just can't help ourselves. One can claim just about anything, and somebody somewhere will be willing to believe it. Over the last several thousand years, no matter where we were or what we were up to, you can be sure there was some heavy-duty believing going on. If you were to board a time machine, spin the dial, and land in any random society of the past, you can be certain that you would find many people who believe in things beyond the normal and natural world that we all see before us. The objects of our belief may vary tremendously, but this common susceptibility, urge or need to believe raises fascinating and important questions: Do we really live in a universe that is overflowing with ghosts, gods, and magical phenomena, as billions of believers declare every day? Or could there be a simpler explanation for most or all of these confident claims that people keep making generation after generation? Are believers describing the real universe or misinterpreting it? One thing is certain, our condition as a believing species remains strong, even in the age of science. According to a Gallup poll, 75 percent of American adults say they believe in at least one of the common supernatural/paranormal beliefs such as ghosts, Atlantis, ESP, psychics, alien abductions, astrology, and so on.[1] Are they reacting to reality or misreading it? Do we have a fundamental need to believe these extraordinary things, or is this just something we do out of tradition and habit? If for one hundred years no adults taught children (or influenced them by example) to believe in magic/paranormal/supernatural claims, would these things vanish? Or would they pop up independently, created anew over and over again? Finally, is it rude to ask these questions, to challenge unusual beliefs? I hope not, because I have been asking them all my life and the last thing I want to be is rude or mean to people. I am only curious, and I want to know as much as I can about my world and my universe. I don't want to derail or distract from this by spending my time believing in and fashioning my life around things that aren't real. What's wrong with that? I would think that most people don't want to believe in things that aren't true. Most people who hold extraordinary beliefs think they are accurate and would want to know if they are not, right?

The intentions of skeptics are often misunderstood, so to be clear, I am not attacking or disparaging anyone when I promote science and encourage skepticism. I'm doing nothing more than suggesting to people that they think more critically *for their own good*. Nobody really wants to waste their time, their money, possibly their *life* on a bogus belief, right? If so, then I'm here to help. We're all on the same side. It is also important for people to recognize that skepticism is not even necessarily antimagic, anti-UFO, anti-ESP, and so on. If unusual claims about magic, alien visits, or miracles are valid, then science and skepticism are not the enemy of those claims. In fact, they are just the opposite. They would be the ways through which to prove that extraordinary and unusual claims are valid.

If we want to know what is real and what is not, if we want to do our best to avoid squandering time and resources on things that are almost certainly not true, then we must make an effort to question everything and do our best to separate truth from fiction. We can't just take someone's word for it that a spaceship crashed at Roswell in 1947 or that the Bermuda Triangle is a paranormal zone of death that swallows up ships and planes because a bestselling book or television show says so. We need to test and verify. Fortunately, we have science, which does this very well. Don't hesitate to utilize the scientific process when you bump up against weird things in everyday life. It includes the following steps:

- **Ask questions.** This is critical. Don't passively accept what you are told. It's stunning how many people fail to simply ask questions when confronted by an unusual claim. Nothing more than a couple of key questions can derail most invalid claims.
- **Observe.** Look and listen with deliberate effort. When someone says "prayer heals any and all illnesses," for example, observe the world. If you see people get sick and die every day—even though they pray—then maybe there is a problem with this claim.
- **Research.** If you look for it, it's not difficult to find credible information about most claims. Do your own fact-checking. If someone tries to sell you magnetic underwear that cures constipation and adds that it's been written up in scientific journals, check to see that it has.
- **Experiment.** Can you think of an experiment to test a particular claim before accepting it? Has anyone else conducted an experiment to test it? If not, why not? If astrology is tempting you, for example, have a friend give you random horoscopes of various signs every day for a month. Log how accurate the horoscope is each day. At the end of the month, see if the horoscopes based on your "real sign" were accurate more often than the others.
- **Share ideas and conclusions with others.** This is a great way to get feedback from people who may know more than you about a given claim. The more good information, the better. Remember the goal is not to debunk or discredit. The goal is to get to the truth, whatever it may be.

The scientific process is not perfect, but nothing else comes close. It doesn't matter who you are, what you are, or where you are. Age, gender, income, educational achievements, nationality, hair color, and shoe size are all irrelevant. The scientific process just works. If you want to be better than most at spotting nonsense and dodging delusions, then step number one is to think! Think like a scientist and become a good skeptic. Go on the offensive when someone tries to sell you a bizarre belief. This does not mean that you must reflexively jump into angry opposition of any claim that hints of magical, supernatural, or paranormal belief. Being a skeptic is not like joining a club or political party and vowing to forever uphold a list of agreed-upon beliefs and nonbeliefs. Good skeptics change their minds, according to the best evidence available. There is just one thing to be loyal to here: reality.

Some people argue that there must be *something* going on because every known society, without exception, has had high levels of belief in various things that can be described as supernatural or paranormal. How could so many people, today and throughout history, be so wrong about so much? With all this smoke, there must be a few fires somewhere, right? Maybe. But a good skeptic needs to see some actual flames. The smoke of belief is not enough. Who cares if a lot of people believe? We need something more than a majority vote before accepting an extraordinary claim. Popularity has never been a reliable measure of truth and reality. Look back on our shared past. How many times have most of the people been dead wrong about something important in a given society? Consider religion. From at least the earliest beginnings of civilization to the present, millions of unique and contradictory gods have come and gone. Based on their biographies alone, they cannot possibly all exist. Logic and basic math make it clear that, at the minimum, most people have been wrong about their gods for most of history. This simple observation alone shows that we are prone to creating the magical and the supernatural out of thin air and then believing in it with total conviction. Believing or not believing in something doesn't make it true. There were times in our past when probably 100 percent of the people on Earth would have dismissed the idea of continental drift if it had been presented to them. *Entire continents float and move around the Earth on a sea of melted rock? You must be nuts!* But our belief in immovable continents never stopped them from moving, not one inch and not for one second.

Reality, as we have seen over and over, operates independently of our beliefs. This alone should make everyone eager, if not desperate, to be a good skeptic. We know that good and smart people can be wrong no matter how confidently they believe in something, so shouldn't it be common sense to constantly test and reconsider our beliefs? Try to remember always that merely being a human raised in a human society sets you up to believe many things that are probably not true. So always ask yourself the simple questions that matter: *Is this real? Why should I believe it? Could all of these people be wrong?* Anything worth holding onto won't be threatened by such simple questions. Only hollow beliefs tremble when confronted by reason, and only false claims collapse when skeptical thinking is applied.

I travel to many countries and ask people about their various beliefs; in those conversations, I am constantly alerted to a severe misperception about what it means to be a good skeptic. Most people, it seems, think of skepticism as a dull, emotionless, and illogical position that requires one to believe in nothing and deny the presence of any mystery and all unanswered questions. They view it as tragic arrogance. This is all wrong. Any good skeptic who understands and appreciates science knows there is a vast universe around us that is bursting with mystery and endless questions. When a typical believer in supernatural/paranormal claims tells me that I don't know everything so I can't possibly know that something like ghosts or UFO visits won't turn out to be true one day, I cringe and patiently explain that I would never claim such a thing in the first place. Of course weird things that seem unlikely or impossible today might become tomorrow's well-established facts. This is one of the reasons I love science so much: It is the one thing that consistently leads us to stunning discoveries. But it makes sense to withhold belief until we get there.

What is this boundary between the natural/normal and the supernatural/paranormal anyway? Where is it and how can we even say there is such a border when we now have a very good understanding of just how weird and surprising nature can be? Who needs supernatural ghosts when we have the scientific possibility of parallel universes and extradimensional beings standing right beside us? Who needs the unfulfilled promise of magic for thrills when we have the astonishingly weird but very real physics of quantum entanglement to amaze and melt our brains? Why bother with ESP and psychics when applied brain science can actually read emotions, expose lies, and move matter with thoughts? All supernatural/paranormal beliefs are one of two things: They are either wrong or simply have not been discovered and confirmed yet. No good skeptic is opposed to everything that may be described as supernatural/paranormal in an automatic sense because we know that some of them may well turn out to be true one day. Most are doubtful, some are about as close to impossible as is possible. But which ones might be right? Can we decide such a thing? Which is more likely to be true: demons or ghosts? Alien abductions or Atlantis? Most alternative medicines or most miracles? How can I possibly choose to believe in any of these things until the scientific process confirms them? I can't see the future. It's best to stay humble, keep an open mind, and withhold belief in weird, unusual, important, and extraordinary things until somebody proves them to be true.

UFOS

Oh, if only this one were true! If I were ever to let desire, bias, emotion, and fantasy bully my skepticism into submission, this would probably be the first unproven extraordinary claim I would surrender to. At some point during most of my lectures and interviews, I make the point that it's okay to hope and dream. There is nothing contradictory or hypocritical about having a wild imagination *and* being a good skeptic. So long as one knows were knowledge ends and hope begins, there is no problem. I have no problem, for example, admitting that I have both the mind of a UFO skeptic and the heart of a UFO believer. For reasons I will share shortly, I don't think aliens are routinely buzzing around our planet. However, I freely accept the possibility that they *could* be. It does not embarrass me to confess that I wish they were. The chance that there could be intelligent life somewhere else in the universe thrills me. Contact with intelligent extraterrestrial beings, or at least discovering that they really are there, would literally be a dream come true for me. But I do my best to think like a good scientist and process claims like a good skeptic should. So before I accept UFO belief—the claim that aliens are already here—it needs to survive a few simple questions. Unfortunately, it can't. But my enthusiasm for the possibility remains.

Not everyone shares my perspective on this. According to a Gallup poll, nearly a quarter of American adults believe that "extraterrestrial beings have visited Earth at some time in the past."[2] A joint study by the National Council on Science and Technology and the National Institute of Statistics and Geography looked at this belief and concluded that as much as one-third of all American adults think aliens are here right now.[3]

Because I am interested in space, find science fascinating, and love to wonder about alien life, I follow NASA (National Aeronautics and Space Administration) missions, the efforts of astrobiologists (scientists who work on questions about what extraterrestrials might be like and where they might be if they exist), and the work of SETI (Search for Extraterrestrial Intelligence). In 2012, after recording a guest segment for SETI's radio show about the intersection of fictional zombies and real science, I pleaded for a personal tour of the SETI Institute and got it. The Allen Telescope Array is not at the main building in Mountain View, California, of course, and it was a Saturday so the offices were mostly vacant, but for a fan like me that didn't matter. I was thrilled simply to be around and peek inside the empty offices of Frank Drake (author of the famed Drake equation that calculates an estimate for the number of extraterrestrial civilizations) and Jill Tarter (longtime SETI scientist and inspiration for the character Ellie, played by Jodie Foster in the movie *Contact*). *Wow*, I thought to myself, *this building is ground zero for the attempt to answer one of the most profound questions ever posed by our species: Are we alone?* Let's just say most six year olds at a Disney theme park show more restraint and dignity than I did that day. My hosts, Seth Shostak and Barbara Vance, were exceedingly gracious and both patiently answered every nerdy question I threw at them.

Near the end of my tour, I noticed an interesting item on a shelf in Shostak's office. It was a fake alien in a specimen jar. I was happy to see that. It seems like maybe he too has the mind of skeptic but the heart of a believer. Shostak is the senior astronomer at SETI and author of the excellent book *Confessions of an Alien Hunter*. Over the years, I've attended his lectures, read his books, interviewed him, and been interviewed by him. I am certain he does not share the UFO belief that is so popular with millions of others. He may have dedicated his life to the search for aliens, but he remains a scientist and a skeptic. Shostak knows that since the hard evidence isn't there, it makes no sense to pretend to know they exist. But he still enjoys the *idea* of aliens being real and even being here among us already. He's a sci-fi film fan and not above keeping a rubber alien around for kicks. I have a few aliens in my life, too. One is perched on the dresser in my bedroom right now. The point is, UFO believers should understand that they don't have to give up the attraction or excitement they may feel about the possibility of extraterrestrial life existing. Keep the enthusiasm, but consider trading in the hollow belief and unfounded certainty for solid science and sensible skepticism.

The key problems with UFO belief are easy to identify. First, of course, is the *total absence of good evidence*. After decades, some say centuries and even millennia, of claims about aliens flying around in our atmosphere, there is *nothing* substantial to show for it. No one has ever produced any evidence that has been scientifically confirmed to be extraterrestrial in origin. All we have are stories, photos, and videos that prove nothing. Eyewitness accounts are more plentiful than camera shots (Who *hasn't* seen an unidentified flying object?) but are even less convincing. This is a very good place to again bring up that old axiom popularized by the late Carl Sagan: "Extraordinary claims require extraordinary evidence." Certainly the claim that alien space-ships are soaring close above our heads qualifies as extraordinary. So where's the extraordinary evidence?

Another problem with UFO belief is that we can't trust our eyes or our memory. As covered in chapter 2, there are standard challenges that come with being human. We don't really *see* what we look at. Instead our brain *tells us* what we see, and it doesn't give us the complete and accurate picture. And never forget that we don't really remember things we saw or experienced in perfect replay fashion. Instead, our brain *tells us a story* about what we saw or what happened, based on incomplete images and bits of information. And, like all stories, memories change and memories aren't necessarily faithful to the true past. All of this is not disputed or controversial in any way. These are basic things we have discovered about ourselves, and they carry huge implications for UFO claims.

How can we trust the eyewitness account of a spaceship sighting when we know eyewitness accounts can't be trusted? How can we trust someone's memory of seeing a spaceship when we know that memo-ries can't be trusted? Once we consider how unreliable our brains can be, it's plain old common sense to conclude that hard evidence— something more than stories from people—is necessary to make the case for UFOs. We need something tangible, like a piece of an extra-terrestrial spaceship that can be analyzed and passed around for our best scientists and engineers to study. Short of that—or an alien pilot, dead or alive—UFO claims are nothing more than high-altitude ghost stories. I certainly wouldn't discourage anyone from investigating specific claims that seem enticing. But until someone produces evidence, I suggest that our time is better spent doing science or at least supporting the efforts of NASA, astrobiologists, professional and amateur astronomers, and SETI. They channel curiosity and hope into scientific activity. And that's the way to do it.

ALTERNATIVE MEDICINE

Alternative medicine is one of those topics that can get skeptics into trouble. It's extremely popular, and many of those who believe in it can be very loyal and protective of it. But fear of controversy is not a good enough reason to keep quiet, not when so many people are ripped off and harmed every day because they believed in unproven medicines and treatments. Anyone who understands the very serious problems with alternative medicine and also cares about their fellow humans has no choice but to speak out. Let's take a quick glance at some of the ways in which this stuff can go terribly wrong and even result in death:

- Helena Rose Kolitwenzew, an eight-year-old diabetic, died after her mother stopped giving her insulin on the advice of alternative-medicine salesman Laurence Perry. Before seeing Perry, the mother had tried many alternative-medicine treatments for the child, including acupuncture and shark-embryo injections. According to a court transcript, the mother said Perry presented himself like a real doctor. He wore a white lab coat, had "medical instruments" in his office, and told her that he consulted the government about viruses. After the child's death, a North Carolina jury found Perry guilty of involuntary manslaughter and practicing medicine without a license.[4]

- A disturbing Harvard study found that an estimated 365,000 people with HIV died prematurely between the years 2000 and 2005 in South Africa because of government decisions to promote alternative medicines over evidence-based medicine.[5]
- A study of Africans who suffered rapid kidney failure found that a high percentage of them developed this serious health problem because they had used harmful alternative medicines.[6]
- Researchers looked at Pakistani women who had been diagnosed with lumps in their breasts but delayed seeking recommended follow-up care. They found that 34 percent of these women failed to get proper help because they chose to rely on alternative treatments, primarily homeopathic medicine and "spiritual therapy."[7] Obviously delays like this can lead to complications and possible death.

Alternative medicine is far from safe, no matter what those who sell it or defend it say. To be fair, weak skepticism is not the only problem. In many cases, particularly in the developing world, folk remedies and traditional treatments may be all that is affordable or accessible. But the fact remains, alternative medicine does harm people.

It is important to keep in mind that this problem of misplaced trust has relatively little to do with intelligence or education. Just about anyone can fall for the unproven claims of complementary and alternative medicine. The best way to keep bogus beliefs at bay is to think like a scientist and be consistently vigilant with your skepticism. Being smart in a general sense is not enough. Nobody thought Apple owner and tech genius Steve Jobs was a dim guy, yet it seems he may have lost his life by relying on alternative medicine to treat his pancreatic cancer. Jobs shared his regret for wasting critical time on things like hydrotherapy, acupuncture, herbal remedies, and even a psychic.[8] Close friend David Kelly said in a 2013 60 Minutes interview that Jobs told him a short time before dying that bad decisions about treatments had doomed him and that he should have trusted in medical science.[9] Biographer Walter Issacson said Jobs had expressed regret in his final days to him as well.[10] The lesson is clear: It doesn't matter how smart, cool, popular, or rich you are. Alternative medicine can seduce anyone who isn't a good skeptic.

One of the bestselling complementary and alternative medicines today is homeopathy. I researched this one for a previous book and was surprised to discover how many people say they believe in it and use it despite knowing little to nothing about what it is or how it is supposed to work. They seem to like the packaging or hear a story from a friend that it works, so they buy it. But slick marketing and word-of-mouth tales should never be the basis for one's personal healthcare decisions.

Homeopathic medicine traces back to a late-eighteenth-century German doctor named Samuel Hahnemann. His "medicine" is very easy to explain because it's water. And when I say water, I mean *water*. Really, there is nothing else in it. I have seen some homeopathic medicines in pill form, but there's nothing significant in those, either. The basic claim of homeopathic medicine is that water can "remember" an active ingredient in the original brew and that—contrary to logic—the more you dilute the solution, the more potent it becomes for treating diseases. Most homeopathic remedies are diluted to such extremes that there is *nothing left* of the original active ingredient! Seriously, I'm not making this up. A typical dose is so diluted that a person would have to drink an estimated 25 metric tons of it in order to have even the slimmest chance of consuming just *one molecule* of the original ingredient.[11]

Even if such outrageous dilutions were not an issue, the logic behind the selection of the active ingredients in the first place is questionable enough. Hahnemann believed that a substance that caused symptoms similar to those caused by a particular disease would help the body cure that disease. So, if a patient suffers from nausea, you should give him a small dose of something else that causes nausea, just enough to trigger his body's defense against nausea. ("Small dose" here means *nothing*, of course.) This is the "like cures like" claim of homeopathy. On the surface, it seems similar to the principle behind vaccines, which do work, but this is different. Vaccines are designed to aid the body in being able to target the disease agent and not just the symptoms of a disease. Vaccines also contain active ingredients, and they have been proven effective by saving hundreds of millions of lives. Finally, the claim that homeopathic water "remembers" the medicine that has been removed from it contradicts what is currently known about water. It's an extraordinary claim, and the entire fields of chemistry and physics would love to see proof for it. So far, however, it's nothing more than a belief.

My local pharmacy has a large display of these products. I've seen them featured prominently in drug stores in the Caribbean and Europe, too. Homeopathic water is touted to be useful in treating virtually anything. I've seen little bottles of it designated for postpartum depression, adrenal fatigue, influenza, anxiety, malaria, Attention Deficit Disorder, psoriasis, joint sprains, depression, migraines, cuts, insomnia, and many more ailments. For all its faults, homeopathy appears to be too scientific for some. I recently found a bizarre display of a similar magic-water product in a major brand-name grocery store in California. The claims for it were astonishing and went beyond even what most homeopathic medicine sellers promise. According to the signage, "doubtfulness," "loneliness," and "low confidence" could be cured by selecting the proper vial and consuming it as instructed. I mentioned this while speaking to a psychology class at Dartmouth University a few weeks later, and one student said he knew another medicine just like it: "It's called beer," he said. But beer is very different, of course, because it contains active ingredients and it works.

To be fair, I suppose I should mention that the widely touted claim that homeopathic medicine is safer than evidence-based medicine is absolutely true. Of course it is, because it's water! When a "medicine" contains nothing but water, the risks of side effects or physical addiction are nonexistent. Unfortunately, of course, it doesn't cure anything, either. Sadly, *belief* in homeopathy does hurt and kill people. In Australia, a homeopathic doctor and his wife allowed their infant daughter to die a slow and painful death because they trusted in alternative medicine over evidence-based medicine. Thomas Sam, a "practicing homeopath," was advised to seek proper care for the baby's severe skin infection but refused. The parents' faith in this alternative medicine was so great that they refused to waver, even as they watched their child's hair turn white, body shrink, skin bleed, and corneas melt. The baby was in constant agony and often screaming, according to testimony during the parents' criminal trial. After suffering for months, she finally died. Both parents—college graduates, by the way—were found guilty for their negligence. The father was sentenced to at least six years in prison, with a maximum sentence of eight years. The wife, Manju Sam, was given a maximum of five years and four months.[12]

In a similar case, a baby died of a subdural hematoma in Japan in 2009 because, unknown to the parents, a midwife allegedly decided on her own to treat it with homeopathic water rather than give the doctor-advised vitamin K supplement, as she had been instructed to do by the parents.[13] Like many complementary and alternative treatments, homeopathic water itself may be safe, but choosing it over real medicine can be costly.

I don't want to go too far here and leave readers with an exaggerated sense of the dangers of alternative medicine. Yes, some of it does cause direct harm and some of it kills. But it's not as if people are dropping like flies in the streets after drinking magic water or getting their backs cracked by quacks in strip malls. Much of it is harmless in a direct sense. There is, however, a serious problem with alternative medicine that goes beyond wasting money and threatening people's health. Alternative medicine is one more thick layer of modern thinking gone bad in which science and skepticism are brushed aside and muted. Unwarranted confidence in alternative medicine is another contributor to the spell of nonsense and irrational fixations that plague our species. By pointing this out, I am not attacking people who believe in alternative medicine; I am attacking the targets of their belief. Consider this description of one alternative medicine:

> Flower essence therapy is a form of vibrational healing, which treats with pure energy to generate changes in the energetic field of the client. Healing with flower essences is an extension of the time-honored tradition of herbal medicine and has been used for over 60 years, successfully returning people and animals to balance and health.[14]

I did not select a fringe treatment to highlight and pick on. I have seen flower-essence-therapy products for sale on the shelves of three large, well-known grocery stores that are less than a ten-minute drive from my house. Chain grocery stores don't give away shelf space to just anyone, so apparently this stuff sells. One of the displays suggests/promises (can't quite tell) that the treatments are appropriate for "fearfulness," "mania," "phobias," and "impatience." I doubt it, but what's the harm other than wasting a few dollars? Sipping, absorbing, or smelling a bit of flower extract isn't going to hurt anyone, right? Here's the problem: Alternative medicine encourages people to trust without evidence and to accept ideas without thinking. It encourages and feeds a culture of gullibility. Show me a person who is capable of believing that flower-essence therapy works as advertised, and I'll show you a person who is capable of believing virtually anything so long as it's packaged and presented in a way that suits his tastes. Belief in alternative medicine is both a symptom and a cause of bad thinking.

But it's natural. The "natural is safer" argument is popular, but that doesn't mean it holds up to a bit of critical thinking. First of all, not every alternative medicine billed as "natural" is safe. Undoubtedly some can be harmful. "Natural" should be recognized as a silly selling point anyway because natural does not equate with "safe." Cone-snail venom is natural, but you don't want any of it in your bloodstream. Cobras are natural, but I wouldn't advise grabbing one. What's more natural than water? Drink too much, however, and even it might kill you.[15]

Real medicine isn't perfect either. A common tactic used by proponents of complementary and alternative medicines is to make the charge that mainstream or evidence-based medicine has many problems, too. It also harms and even kills people. I agree! Many drugs have brutal side effects. Not all doctors are competent. Not every healthcare administrator is ethical. Many times medical science just has no answer for a disease and can't help a patient. It is also true that for many sick people a visit to the doctor's office can be dehumanizing because it feels like a trip to an automobile repair shop. Much of the criticism about mainstream healthcare that comes from fans of alternative medicine is correct. Perhaps because it has devoted so much of its focus on coming up with the things that actually work against injury and disease, medical science has failed in many respects to keep the patient in mind. Somewhere along the way, it seems too many doctors forgot that we are human beings, something more than a collection of organs with glitches in need of repair. We have feelings and fears that doctors don't always consider. There also seems to be too much emphasis on disease treatment or management as opposed to disease prevention. My response to alternative-medicine fans is to agree with them that there are significant problems with mainstream, evidence-based healthcare. But this does not make their case or support their claims. The failures of medical science do not prove that things like homeopathy, reflexology, and ear candling actually work.

Another key problem with alternative medicine stems from problems of interpretation. We can be very bad at identifying causes. For example, many people get sick, go to a doctor for treatment, but also take an alternative medicine at the same time. When they recover, however, they may give no credit to the doctor or prescribed medicine. Nor do they give any consideration to the possibility that it might have been nothing more than time and their own body's recuperative powers that did it. They believe and declare to anyone who will listen that an alternative medicine cured them, despite the fact that they cannot know this to be true.

I have never taken alternative medicines in my life, yet I have recovered from every illness I have ever had so far. Had I taken an alternative medicine during one or more of my illnesses, however, I might have been led to believe that it helped me, even though I couldn't know that for sure. Unfortunately, problems with attributing credit where it is undeserved doesn't stop people from spreading stories about the herbal potion that cured some guy down the street or the touch-therapy session that fixed Uncle Joe's bum knee. This raises another problem: Stories are easy to tell and easy to believe. Please understand and remember that mere stories should never be the basis for deciding on the best medical treatment of injuries or disease. Always think about *sample size*. Has it been scientifically tested on enough people to draw a reasonable conclusion about it? Hearing from a few people here and there who rave about an alternative medicine is just not good enough. Random testimonies won't do, not if you want to be sensible about truth, reality, and your health. There is a very good reason why medical researchers don't find just one or two people who had a positive outcome with a drug and then promptly declare it to be effective and send it out to the marketplace. Testing methods and sample size matter. The only way to be sure about a drug is to submit it to the scientific process. It's not foolproof, but it works better than anything else. What is required in most cases is a study with multiple people receiving either the medicine in question or placebos (inert sugar pills, for example), all distributed randomly, without the researchers or patients knowing who got what until it's over. This way, bias can be removed (hopefully) and it can be determined if the medicine works.

It's always a good idea to consider the placebo effect when hearing about some alternative medicine that a salesperson, friend, or family member is raving about. This strange phenomenon is very real and undoubtedly is responsible for much of medical quackery's success. Some people some of the time can get a positive health benefit from taking a fake medicine (a pill, shot, or treatment that has no medicinal value) instead of real medicine. This is well documented but still is not fully understood. The problem you need to keep in mind, however, is that it's not consistent, and even if there is some positive gain, it might not be enough to get one through an illness safely. So the placebo effect is not something anyone should rely on.

Ultimately, I think the best way to show people what is wrong with alternative medicine is simply to define it. I'm convinced that if more people understood why it is categorized as it is in the first place, then it would not be nearly as popular. Complementary or alternative medicine is really just *unproven medicine*. That's not an unfair criticism; that's just what it is. When an alternative medicine is openly put to the test and confirmed as useful by the scientific process, then it becomes just plain *medicine*. So we really should call complementary or alternative medicine *unproven medicine* and nothing more. By calling it something else, many people are fooled into thinking that it is more credible than it really is. To be clear, "unproven" does not necessarily mean that it doesn't work. I am sure there are some alternative medicines and treatments out there that do work to some significant degree beyond the placebo effect. But the problem is that we can't be sure which ones work and which ones do not until someone uses the scientific method to find out.

GHOSTS

Standing alone in the hallway of the "most haunted house in America," I'm about five minutes away from a freaky experience. The Whaley House is a nineteenth-century Greek-revival mansion that once served as a family home and the court house for Old Town, San Diego. It's been well preserved as a historical site and is a popular point of interest for tourists. Before I began exploring the house, I spoke with a ghost believer outside who said she had a terrifying encounter with former owner Thomas Whaley inside. Dead since 1890, Whaley still roams the property, according to local lore. "There is definitely a presence in that house," she said. "All the sudden, I was freezing cold. I was so scared; I remember closing my eyes and then I knew he was there. I just knew he was going to pass right through me. I was so scared that I couldn't move, even though I knew something was about to happen. And then something went through my hair. I felt it touch my hair. You would never believe how it felt. It was pure death."

"There is definitely a presence in that house," she said again. "The family is there. I think they are attached to the house. You feel a presence. They don't seem to ever leave."

Great story. Extra credit for the dramatic delivery. Good thing I don't believe in ghosts, or I wouldn't go near this house.

The Whaley home is clean and in excellent condition for its age. It's far from the old, dusty, cobweb-draped haunted-house cliché. The ambiance is more Monticello than Amityville, more Brady Bunch than Adams Family. But I'm about to get a lesson in human vulnerability to context and suggestion. I may be a hardcore skeptic, but I'm still human, and the last thing I should have done is listen to a creepy ghost story from some jittery paranormal fan just before entering the "most haunted house in America" all by myself. Something happened to me that day, and I'm sure it had a lot to do with that lady tickling my amygdala (brain's fear center) beforehand. I'll spare you the suspense; I didn't see any ghosts. Mr. Whaley didn't make me cold, touch my hair, or pass through me. And, no, I didn't find any reasons to reconsider my position on the existence of ghosts. But I did *feel a presence*.

The haunted brain. It is a slow day at the Whaley House. Only a few other visitors are downstairs, and upstairs I'm all alone. *Or am I?* In one of the well-preserved bedrooms, I can see old clothes, books, a quill pen, a hairbrush on the dresser, and so on. The presentation is excellent. It looks as if the Whaley couple might round the corner at any moment and tuck themselves into bed. But it was at the children's room where things got weird.

A small doll sits in an old rocking chair. The soulless little demon stares through me with her little, dead, black doll eyes. Forget the house, this damn doll alone could inspire a dozen Stephen King novels. Then, as if on cue, a brief flash of bright light startles me. I spin around immediately, but nothing is there. I'm stumped at first, but then I realize it was probably the Sun's reflection off the window of a passing car outside. Good thing I came up with that possible explanation, or it would have bugged me for the rest of the day. Yes, I admit it; my heart rate is up a bit and I'm extra perky now, thanks to a tiny dose of adrenaline. Continuing on, my mind remains slightly haunted. Staring down a long hallway, I imagine the Whaley family that lived here in the

1800s. I "see" children playing and parents talking, arguing, just living. Then I think about the fact that they are all dead. Okay, I'm really getting my money's worth now. I see dead people—but only in my head. Are they really here? Could there be a family of ghosts drifting around me? Maybe, but I doubt it, because here at the Whaley House, it's just like every other "haunted" place: no good evidence for the claims.

Every sincere story of a ghost encounter likely can be explained as misinterpreted sights and sounds or the weird feelings people have when standard human imagination and irrational fear give them a poke. Billions of my fellow humans may say they know ghosts are real, but not one of them can prove it. Most ghost believers probably don't realize that they are not so different from skeptics when they feel a creepy vibe, sense a presence, or are startled by an unexpected sight or sound. I experienced all of those things within the space of fifteen minutes—despite the fact that I never thought for a second that ghosts are real.

You may be wondering how many people believe in ghosts. The answer? A lot. A Harris Poll study found that 42 percent of American adults think ghosts are real.[16] The percentage of children who believe is probably much higher. According to a Gallup poll, 37 percent of American adults believe that houses can be haunted by ghosts[17] and twenty percent of US adults say they have visited or lived in a haunted house.[18]

I think the most important thing for ghost believers to keep in mind is that the best "evidence" we have for ghosts comes in the form of eyewitness accounts. I won't list the reasons for not trusting eyewitness accounts here. Instead, just consider that many of these reports are about things people heard or saw that they *interpreted* to be a ghost. Remember the woman I spoke with before entering the Whaley House? She never told me that she saw Thomas Whaley or heard his voice. She said she "felt his presence" and "knew it was him." Doesn't that sound like nothing more than imagination run amok? I'm not judging her harshly for having experienced it. This sort of thing is so common that it seems to be a normal human reaction in certain circumstances. After all, nothing more than a flash of reflected light and a creepy doll briefly spooked me once.

In order to be good skeptics who are not easily fooled into believing in things that don't exist, we have to resist applying answers when we don't really have answers. For example, a classic haunted-house experience might involve someone lying in bed and hearing creepy sounds coming from down the hall or from another room. Maybe it's a creak in the wooden floor or the rattle of something that sounds like chains. No doubt, millions of ghost stories have come from such sounds. Never mind that houses, especially old ones, can generate many sounds for many reasons. One mouse scurrying around inside a wall or one drafty room can do it. But guess what? An unidentified sound is an *unidentified sound*. A good skeptic does not pretend to know something she or he does not. The same rule applies with visual experiences. If someone sees a mysterious shadow, a moving wisp of fog, or an unexpected flash of light as I did, "ghost" is not a justified answer to explain it without a lot more evidence to back it up. Many friends and strangers have shared their ghost stories with me, and I would estimate that at least 90 percent of them involve unknown sights or sounds that they *interpreted* to be a ghost. They readily admitted to me that they did not see a distinct humanoid with clearly visible facial features standing or hovering before them and speaking about the hardships of being a ghosts, haunting homes, or whatever. Most reported ghost encounters in real life are nothing like *A Christmas Carol*, in which one clearly sees and carries on conversations with spirits. They are almost always nothing more than unknown phenomenon unjustly defined as ghosts.

I probably should add that I am not antighost. I'm just proreality. As a source of fictional fun and cheap thrills, I have no problem with ghost stories. Within reason, I think it's great to exploit our insecurities and fears about death for a good emotional jolt now and then. Scary stories are near and dear to my heart. I love taking my son to Monsterpalooza, for example. This annual horror-movie and monster-makeup festival in Burbank, California, is a blast. Ghosts are fun. Monsters are cool. I just think we all ought to know the differences between fantasy, reality, and the unknown. One can be a skeptic and a ghost fan, too. Nothing wrong with that. Fear is a terrible, unwanted thing that we avoid at all costs—except when we seek it out. Roller coasters, horror movies, and scary novels frighten us in ways we love. It's even okay to be afraid of the occasional shadow or mysterious bump in the night. That stuff can keep us on our toes and remind us that we are alive and want to stay that way. Just make sure to always do your best to distinguish between truth and fiction, between the known and the unknown.

ALIEN ABDUCTIONS

Have you ever found a bruise on your body and been unable figure out how it happened? If so, how do you know it wasn't from an injury sustained when aliens took you from your bedroom at night to study and experiment on you aboard their spaceship? It's possible. Maybe they wiped the event from your memory. Have you been feeling a little off lately? Does something seem wrong but you can't quite put your finger on it? Perhaps it will help if you visit a hypnotherapist, maybe one who specializes in recovering lost memories of traumatic events such as alien encounters.

After a few sessions of hypnotherapy, you can clearly recall many things about the night aliens took you away to do unspeakable horrors to your body. You feel a bit better now that you know what happened, but you are still haunted by the memories of your abduction.

This may seem like a weird and unlikely process to most people, but it's no joke. Some people seem to believe sincerely that something similar to this happened to them, and they carry around a lot of pain and anxiety because of it. Some of their stories include horrifying experiments that sound a lot like torture and even rape. Others involve the harvesting of eggs or sperm and maybe the implantation of fetuses or alien technological devices. The skeptical view of all this, of course, is that no actual aliens were likely involved and it's more reasonable to think this is probably a collision of imagination, highly questionable mental-health treatment, cultural influence, and the vulnerability of human memory to contamination by suggestion. According to one poll, nearly four million American adults believe they have "encountered bright lights and incurred strange bodily marks indicative of a possible encounter with aliens".[19] If that is anywhere near accurate, it means a *lot of people* think they have been close to or were actually snatched by aliens. On top of that, we have to add all those who believe it has happened to other people.

In fairness to people who claim to have been abducted, I think it is important to point out that mental health, intelligence, and education do not seem to have much, if anything, to do with this belief. It's important to mention this because alien abductees are often the butt of jokes in everything from casual conversations to Hollywood films these days. But be careful. Before you laugh at them, consider the fact that something similar might happen to you. Ever heard of sleep paralysis with hallucinations? This surprisingly common phenomenon may occur in as much as 20 percent of the adult population.[20] Think about this statistic: two out of every ten people. That represents a lot of people. And who knows? You might be next. When we fall asleep, our brains normally will restrict our physical movements so we don't spend the entire night punching ourselves in the head while dreaming. As we wake up, our brains release our bodies from this state of partial paralysis. But sometimes, for some people, the brain "wakes up" before it frees the body. These people find themselves partially awake but unable to move. Add to that an ongoing dream/nightmare or hallucination, and you have the makings of a confusing and potentially disturbing event.

I have a close friend who has had several of these episodes but without the dramatic backdrop of a nightmare or a hallucination. She says she understands very well how one could easily confuse a dream for reality in that state. She also said she has been able to hear sounds and even smell things while unable to move. Fortunately for her, she has had enough of them to know not to panic and just go back to sleep or ride it out until she fully awakes. Clearly, sleep paralysis is a real phenomenon that would seem to set the stage perfectly for an imagined alien-abduction experience in some minority of people. But where do the specific and often-detailed memories of aliens, spaceships, and rude experiments come from? That is easily explained by clarifying how human memory works.

As we saw in chapter 2, memory can fool the best of us. It's just not anything close to a DVR playback system like most people imagine. We remember things by a weird process of *constructing stories* about things that may or may not have happened in the past. Without conscious consent, our brains serve up memories to us that can feel 100 percent accurate and reliable. But, in fact, they have been edited, trimmed down to reflect what is probably important and useful to us in a given moment. Elements may also be added whether or not they happened at another time in our lives, happened to someone else, or never happened to anyone anywhere. Our memories can also be heavily influenced by input from the stories we hear, the books we read, the movies we watch, and so on. Obviously, a therapist who believes in alien abductions and assumes it happened to a client could probably coax and encourage him or her into conjuring up such a memory.

Does any of this prove that no one has ever been abducted by aliens? Of course not. As strange and unlikely as it may be, it's still possible. After all, it's a very big universe. So far as we know, there is nothing impossible about intelligent beings visiting us. And if they did, it's not unreasonable to think they might grab a few of us to study, much like an entomologist might snatch a few ants in the Amazon for close inspection. However, the absence of strong evidence, coupled with what we know about the ways in which people can come up with false memories, makes it clear that this is a conclusion that can't be rationally defended.

Finally, stories of creepy home invasions in the middle of the night by weird nonhuman beings are nothing new. The only thing different today is that they park a spaceship on the lawn. The ghosts and goblins of our past seem to have been updated to reflect our high-tech, space-age times. Abduction skeptic Carl Sagan recognized this: "We had demons from ancient Greece, gods who came down and mated with humans, incubi, succubi in the Middle Ages who sexually abused people while they were sleeping. We had fairies. And now we have aliens. To me, it all seems very familiar."[21]

BIGFOOT AND CRYPTOZOOLOGY

Cryptozoology is the "science" of mythical or undiscovered animals such as dragons, the Loch Ness monster, Yeti (the abominable snowman), and Bigfoot. But cryptozoology is better thought of as pseudoscience rather than science because it operates in reverse of the scientific process. Its fans seem to start with a firm conclusion (these animals definitely do exist) and then go about trying to come up with evidence and arguments to back it up. Science doesn't work that way. It requires the proof to come before the conclusion. Scientists may have wild hunches or flimsy leads that direct them to search for strange, undiscovered animals somewhere—there's nothing wrong with that—but they don't declare to *know* that an unknown animal exists before they can *prove* that it does.

Despite the failure of Bigfoot enthusiasts to prove their claim after all these years, 16 percent of American adults say that the creature is "absolutely" or "probably" real.[22] Before we analyze Bigfoot belief, however, I want to point out that this is yet another extraordinary belief that is very, very unlikely to be true—*but I wish it were*. It would make my day to hear news of one captured alive or a recovered body. It may not win me over, but it does bring a smile to my face when I watch "real footage" of Bigfoot or see the plaster cast of a giant footprint. I love apes and I love the idea of an unknown ape species living right under our noses all this time. You have to understand, I'm a guy who obsessed over the original *Planet of the Apes* films in childhood. When I conducted long interviews with Donald Johansen, discoverer of the famous "Lucy" fossils, and Jane Goodall, the great primatologist, for feature stories, it took a Herculean effort on my part to be professional and stop asking them questions after an hour or so. I even managed to talk my wife into letting me display my museum-quality replica *Australopithecus afarensis* and *Homo erectus* skulls in our living room. I feel closer to a couple of the bonobos at the San Diego Zoo than I do to half the humans I know. Come to think of it, I'm pretty sure that I would be happier and more excited than most cryptozoologists if Bigfoot were found. I certainly wouldn't be opposed to good evidence, attempt to suppress or deny it, or feel awkward about it, anyway. Remember, good skeptics don't declare absolute knowledge about the *nonexistence* of flying saucers, ghosts, and so on. We simply point out problems with the claims and ask for good evidence. When it doesn't come, we do the sensible thing and conclude that these things are unlikely to be true so they aren't worth believing in. But we stop short of claiming to know what the world and the universe do *not* contain, because it's a big world and a big universe, so we can't be sure. Finding Bigfoot would be great for science because it would require an update to our current knowledge of primate evolution. We would have a new species to welcome into the family. Believers in these kinds of things often think that skeptics are against them and their claims. But this is not true. We simply know better than to pretend to know things we don't know—even when we might love for them to be true.

So what are some of the problems with the claim that there is a giant primate running around in the Pacific Northwest and/or southern swamps of America? The biggest challenge for Bigfoot believers is simple: no body. Skeptics say "show me the body," just one. But it hasn't happened yet, and that's a significant reason for doubt because it is extremely unlikely that after all this time not one camper, forest ranger, hiker, runner, logger, mountain biker, fisherman, or hunter would have stumbled upon a Bigfoot body that had died of old age or injury. All it would take is one body. Just one ten-foot tall bipedal ape in North America and Bigfoot is confirmed. Actually, it wouldn't even take that much. Biological anthropologists are so good that they could prove it with only a few key bones. After all, they were able to figure out that *Gigantopithecus*, a 1,200-pound and nine-foot-tall ape, roamed the forests of Asia hundreds of thousands of years ago. And they didn't even need a complete body to do it. Fossilized teeth and jawbones were enough. So all we likely would need to confirm Bigfoot would be *part* of a specimen. But we have nothing. All these years, and not even one jawbone turns up? No teeth? Not one unusually long primate femur bone? What are we supposed to believe, that the Bigfoot clans bury their dead in secret, unmarked graves out of sight of outdoor enthusiasts and forever beyond the reach of expanding suburbs?

Another critical problem for the Bigfoot claim is that there can't just be one out there. Very few people think about this, but for Bigfoot creatures to endure all these years, there would have to be many of them out in the woods eluding people year after year. I raised this point with an anthropologist friend of mine who lives in Oregon (Bigfoot country). He guesses there would have to be at least five hundred Bigfoot individuals in a given area to maintain a genetically viable population capable of surviving long term. Of course, hundreds of creatures would make it even more likely that remains would be found. But they haven't been. Disappointing, yes, but reality does not always cater to our desires. It just is what it is.

I know what you are thinking. *What about that famous Bigfoot film? And what about all those plaster casts of giant footprints? Isn't that stuff proof?* No, it's not because we have very good explanations for those things that don't require a population of gigantic apes. First, let's take care of the Patterson-Gimlin Bigfoot film. Shot by Robert Patterson and Bob Gimlin in Northern California in 1967, this brief bit of footage has impressed millions of people around the world, undoubtedly converting many into believers. I have no idea why, because it's terrible. (If you are not familiar with it or haven't seen it recently, search for "Patterson Bigfoot film" online and judge for yourself). I'm no primatologist, but the "creature" seems all wrong to me. It does not look and move like a wild animal. But it does look and move like a human in an ape suit. Furthermore, the way in which it parades by Patterson and Gimlin and then gives the men a dramatic money shot by glancing back at the camera before vanishing back into the woods seems too convenient. Even more difficult to believe is this little detail: Patterson told people in advance that he was going out that day to find and film Bigfoot. Ask yourself, what are the odds that he would have been that lucky? Maybe a pudgy, flat-footed, pear-shaped Bigfoot really did accommodate the two men that day, but I doubt it. It has all the look and feel of a staged hoax, and that ought to be everyone's first guess because we know hoaxes happen. But let's not rely on my opinion here. The man who seems to have worn the suit, Bob Heironimos, came forward and confessed![23] Greg Long's book, *The Making of Bigfoot*, includes that and more. It lays out the whole story in detail and includes information about Patterson buying a gorilla suit from Morris Costumes in 1967. The owner of that business, former magician Phillip Morris, said Patterson told him he planned to "have some fun with the suit."[24] Seems he did, indeed.

Evidence in the form of plaster casts of Bigfoot prints is even less impressive than the film. Faking giant footprints was something of a cottage industry in the Pacific Northwest in the 1970s. For a while, it seemed like everyone was doing it. But the first print-maker may have been Ray L. Wallace, a road-construction worker. When he died in 2002, his son, Michael Wallace, told the *New York Times* that his father was an enthusiastic prankster who made giant wooden feet and began stomping around in the forest, making prints, from as early as 1958. He said his father never intended it to be a complex plot, nor did he imagine that it would grow into the widespread phenomenon that it did. He only meant to play a trick on a few local people for a laugh.[25]

Science finds the real monsters. The good news for veteran or aspiring cryptozoologists is that they can do what they love—think about, research, and search for exotic new life-forms—*within* science instead of outside of it. If the chance to find new creatures sounds exciting to you, then science is the ticket. Take your pick: biology, zoology, microbiology, entomology, astrobiology, and more. These scientific disciplines are not only *open* to new discoveries, they, unlike cryptozoology, actually *make* them on a regular basis!

Many people are not aware of it, but science has a long, long way to go when it comes to finding and cataloging the life-forms we share this planet with. There is no doubt that the Amazon rainforest, the hills and mountains of Asia, the African bush, and other more-remote locations are still hiding many surprises from us. Even more promising is the ocean. Scientists can hardly pull up a water sample without finding something new in it. Maverick scientist and visionary Craig Venter of San Diego has had teams trolling the ocean for nine years in search of new life. So far, he and his team have discovered *hundreds of thousands* of new species and *sixty million* previously unknown genes.[26]

Forget life on the planet for the moment, we haven't even identified the life on us! There is still a tremendous amount of work to be done to discover and understand the microbes that share our bodies with us. Never make the mistake of thinking that the only interesting, important, or scary creatures are big. Sure, a ten-foot ape would be impressive, but have you ever seen images of the little beasts crawling around on your skin and inside your home right now? They look like monsters to me. Have you learned about some of the amazing and bizarre things that microbes can do and the surprising places they can live? No doubt, there are many millions of real monsters awaiting discovery right now, and science is the route to finding them. Popular Harvard scientist Edward O. Wilson is in awe of the still-mysterious microbial frontier:

> Bacteria, protistans, nematodes, mites and other minute creatures swarm around us, an animate matrix that binds Earth's surface. They are objects of endless study and admiration, if we are willing to sweep our vision down from the world lined by the horizon to include the world an arm's length away. A lifetime can be spent in a Megellanic voyage around the trunk of a single tree.[27]

One does not even have to limit the search for new life to this planet. Astrobiology is a respectable and growing branch of science that treats the entire universe as a target of opportunity. Astrobiologists do not claim to know that extraterrestrial life exists, of course, only that it's certainly possible and well worth thinking about and looking for. Much of their work involves the search for and study of weird creatures here on Earth that live in extremely hot, extremely cold, or extremely acidic environments, the kinds of places once thought to be impossibly hostile to life. That line of work has to be more exciting and rewarding than looking for the Loch Ness monster, Bigfoot, or other mythical monsters that are very, very unlikely ever to be found.

PSYCHICS

Tens of millions of Americans and billions worldwide believe in strange mental powers that the scientific process has been unable to confirm. According to a Gallup poll, 41 percent of American adults believe in ESP, 31 percent believe in telepathy, and 21 percent believe that some people can communicate with the dead.[28] They embrace these claims despite the failure of everyone who has ever tried to prove any of it. Apparently, people are sufficiently impressed by the work of professional psychics (mind readers with knowledge of the future) and mediums (live people who talk with dead people). But there are very simple explanations for how psychics and mediums are able to impress so many people. For example, a *cold reading* can exploit natural human biases and lead some people to believe that there is something supernatural going on when there is not. I was so intrigued by this that I decided to give it a try.

Although I had read enough about cold readings to know how they work, I was still hesitant. The method requires you to use educated guesses about a person for you to convince her that you are peering into her mind. At the very least, it takes a lot of nerve to sit face-to-face with someone and pretend to read his or her mind. It's useful to know many trends and traits that are common to various people, genders, religions, nationalities, ages, and so on. It also helps to be able to think fast and make immediate adjustments based on real-time reactions and feedback from the subject. Researching cold readings is one thing. Actually doing one, however, is something entirely different. But I wanted to at least try, so I plunged in.

My first "client" is a woman in her midthirties. She's bright, attractive, and, most importantly, tells me she thinks there is something to psychic powers. She is predisposed to believe me. That's about 80 percent of the battle won right there because prior beliefs and expectations are known to be powerful influences on how we interpret our experiences. My confidence soars.

She stares, waiting for me to start. I speak slowly and pause a lot, hoping this makes me seem sure of myself. I spend a few moments blabbing about nothing, mostly empty words about me "feeling" her thoughts and how I need her to relax and open up to me. I begin the actual cold reading by peppering her with several rapid-fire statements and questions. While doing this, I pay close attention to her eyes, facial expressions, and body language. I look for any reaction to each thing I say. I notice she is not wearing a wedding ring. Sure enough, she lights up with heightened interest when I blurt out something about frustration with her love life. She's in her thirties, and I guessed that marriage might be on her mind. Her reaction indicates to me that I should zero in on this subject, so I pretend to know all about boyfriends who have disappointed her in the past, as if that makes her unique. Once I feel I've gone as far as I can with made-up predictions and crude generalizations that would fit many women, I move on to money. I tell her she's not greedy but could use a little more. Then I add in something about just wanting to be a good person and lead a happy life. Overall, I'm feeling pretty good about my performance. But is she buying it? Does she think I'm reading her mind and can see the future? Maybe. I'm not sure. I keep going. I try to ask questions and make statements that have a good chance of feeling personal and relevant to her based on her gender, age, nationality, religion, and so on. I mix in a few flattering comments. It's surprising how many clues one can pick up when "reading the mind" of a person sitting two feet in front of you. She smiles when I say something that hits and she appears neutral or uncomfortable when I clearly miss the mark. It's as easy as that. I end by giving her a cliché-filled pep talk: Your future is bright; try your best; never give up.

When I finish, she thanks me but leaves abruptly. That's it; show's over. I'm left deflated and disappointed. I conclude that I somehow blew it. Maybe cold readings are a lot more difficult than I thought. Somewhere Sylvia Browne and a million spirits are laughing at me.

But I see the woman a couple of days later and she approaches me.

"How did you do that the other day?" she asks. "It was amazing. How long have you been a psychic? You knew so much about me."

Wow. Clearly I did better than I thought. She was *completely* convinced. Of course, I quickly explained to her that I did not have any special mind-reading powers and that I was attempting a cold reading just to see if I could do it. I apologized and said I would have explained it that day but didn't think she believed me so it wasn't necessary. The woman was a bit unsettled at first, but as I told her more about how it works and explained confirmation bias, she seemed grateful for the knowledge. The lesson I took away from that experience is that it doesn't take much to convince someone. If I could do this well on my first try, it's no wonder that some people who spend years perfecting this routine are able to convince many people and get rich.

This was a stunning realization for me. I had stumbled through a performance that was mediocre at best, but it turned out to be more than good enough. How did I do so well? She believed in psychics to some degree beforehand so was predisposed to accept my act. She also was a human with a human brain. This means, of course, as we saw in chapter 2, that she was vulnerable to confirmation bias. Like all victims of cold readings, she probably clung to and remembered everything I said about her that was correct or close enough because it reinforced her belief in psychics. But she probably missed or soon forgot everything I said that was clearly wrong because my errors did not confirm her bias. Then her memory probably helped me out by exaggerating how good I was every time she recalled our session.

It's important to understand confirmation bias and be alert to it all the time because it can trip up people not only during a psychic reading but in the context of many other extraordinary claims as well. When we selectively embrace input that matches a conclusion we have already made, virtually anything can seem real and reasonable, regardless of how fake and unreasonable it is. It's like throwing a thousand darts in the dark and then turning on the lights to find that you hit the bull's-eye ten times. Hitting a target ten times in the dark might seem very impressive, but only if you ignore, forget, or never see those 990 darts that missed. When a psychic says a hundred things, we can't allow ourselves to take note only of the correct things and forget all about the incorrect things he says. And we can't give credit for vague statements and questions. We need to keep score in order to determine if there is anything to this or if it's just a bunch of semieducated guesses designed to fool people. And don't forget sample size. We can't restrict ourselves to considering only one or two individual readings when assessing a psychic. If a psychic does hundreds of readings, for example, it is to be expected that at least a few might go exceptionally well and appear to be eerily accurate. However, placed in proper context with all the readings that go poorly, it's not impressive.

I have a friend who was good enough at psychic readings to make money at it. He wasn't a cold-hearted con artist, nor was he doing it as a noble experiment toward a better understanding of the human mind. He was self-deluded, fully convinced that he possessed special powers. But even as the cash rolled in and clients raved about his abilities, he began to have doubts. He noticed that he got credit for making guesses about people, things anyone could come up with. Eventually, he stopped believing not only in his own psychic powers but in all supernatural and paranormal claims. Today he is a good skeptic and big fan of the scientific process. He stressed to me how easy it was to fool people, including himself.

Operating much like psychics, mediums claim to carry on two-way conversations with dead people. Many psychics are mediums as well, which makes sense because it's the same game. They blast out a bunch of guesses, questions, and platitudes about a dead person and then watch to see what gets a reaction from the surviving spouse, family member, or friend. *He's here with us now and he tells me to tell you that he loves you and is proud of you for doing your best in life. He also said he likes that you keep a photo of him. He says keep working hard and believe in yourself.* A good skeptic sees right through this. For some grieving people, however, it can be irresistible. It's understandable to want to communicate or connect with someone who is gone, but that desire should not override the ability we all possess to think our way through things like this. No one should allow himself or herself to be emotionally exploited and toyed with, even if it does feel good for a fleeting moment.

The bottom line with psychics and mediums is that their claims can be tested. There is a way to confirm once and for all if people can peer into the mind of a stranger, living or dead, and know intimate details about his past, present, and future. If psychics are interested, science can confirm their abilities to the world. All they have to do is submit to the process. It is telling that very few do. If psychics and mediums are not sufficiently motivated by the prospect of scientific confirmation for the sake of truth, one might think that money would do the trick. The James Randi Educational Foundation has a standing one-million-dollar prize available to anyone who can demonstrate a paranormal/supernatural ability under the conditions of a credible scientific test that she or he can design. But for reasons one can easily imagine, today's best-known psychics and mediums don't even try. They declare that they can read minds and glean knowledge from dead people but then leave a million dollars on the table.

Hopefully you will be too good a skeptic to fall for the games psychics and mediums play. But you are bound to know someone who believes. Tell that person about me. Tell her the story about the guy who, on his first try, was able to fumble and stumble his way through a short cold reading and convince a smart adult that he was a mind reader. If that guy can do it, anyone can, so keep your guard up.

THE ROSWELL UFO CRASH

Without a doubt, the best UFO story of all time is the one about an extraterrestrial spaceship crashing in the desert near the tiny town of Roswell, New Mexico, in the summer of 1947. This was no mere sighting of a strange light in the sky. No, this time we had confirmation from the US military, recovered wreckage, and maybe even the dead bodies of aliens to prove it. Roswell is the Holy Grail of UFO belief. But there's one catch: it's almost certainly not true.

Don't get me wrong, I love the "Roswell Incident." But I know better than to accept it because I have studied the claims and the facts, spoken with a test pilot who may have inadvertently contributed to the myth, and considered how the tale works so well with the human propensity to believe without evidence. That is not to say, however, that there isn't a true story to be told here. Roswell believers are right about a few key elements. Something unusual did crash in the desert. Strange wreckage was recovered. And there really was a government cover-up about it.

It all started when local rancher Mack Brazel found debris in the desert outside of Roswell in June 1947. He assumed it was just trash and moved on. (Not the reaction one would expect if it was a crashed flying vehicle of any kind.) Several days later, a private pilot named Ken Arnold reported seeing weird objects while flying in the Pacific Northwest, far from Roswell. He described the objects as flat and shaped something like boomerangs, but the term "flying saucer" was widely reported in the media and that's what stuck with the US public. It is interesting to think that more accurate reporting of Arnold's sighting would have led to the "flying boomerang" craze of the 1950s instead of flying saucers and all those black-and-white sci-fi B movies of the era would have depicted flying boomerangs terrorizing the Earth. In the wake of the Arnold story, thousands of people across America made UFO reports in the following months and years. The UFO era had taken flight, and it likely influenced the events at Roswell.

Some three weeks after he found the wreckage, Brazel drove into town, where he might have heard about Arnold's UFO sighting or other "flying saucer" stories. Regardless, for one reason or another, Brazel now felt that what he found might be important, so he reported it to Roswell's sheriff, who in turn alerted the nearby Roswell Army Airfield. Personnel from the base recovered the materials and then the story really took off because an enthusiastic public-information officer issued a press release claiming that a "flying disc" had been recovered. Soon after that, the *Roswell Daily Record* newspaper carried a lead story about it with this headline: "RAAF Captures Flying Saucer in Roswell Region."[29] It's worth noting that in those days "flying saucer" or "flying disc" didn't necessarily mean alien spacecraft to people the way it does today. So maybe the press release wasn't quite as overreaching then as it seems to us today. A very important point in all of this is that immediate descriptions of the material by Brazel, the original discoverer, seem to make it clear that this was absolutely *not* the remains of spaceship from an advanced extraterrestrial civilization. The following excerpts are from a front-page report printed on July 9, 1947, in the *Roswell Daily Record* on page 1, just one day after the original "flying disc" story. *(Bold added for emphasis.)*

- *"Wilcox* [Roswell sheriff] *got in touch with the Roswell Army Air Field and Maj. Jesse A. Marcel and a man in plain clothes accompanied him home, where they picked up the rest of the pieces of the "disk" and went to his home to try to reconstruct it."*
- *"According to Brazel they simply could not reconstruct it at all.* **They tried to make a kite out of it,** *but could not do that and could not find any way to put it back together so that it could fit."*
- *"Brazel said that he did not see it fall from the sky and did not see it before it was torn up, so he did not know the size or shape it might have been, but* **he thought it might have been about as large as a table top.** *The balloon which held it up, if that was how it worked, must have been about 12 feet long, he felt, measuring the distance by the size of the room in which he sat.* **The rubber was smoky gray in color** *and scattered over an area about 200 yards in diameter."*
- *"When the debris was gathered up the* **tinfoil, paper, tape, and sticks** *made a bundle about three feet long and 7 or 8 inches thick, while the rubber made a bundle about 18 or 20 inches long and about 8 inches thick. In all,* **he estimated, the entire lot would have weighed maybe five pounds."*
- *"There was* **no sign of any metal in the area** *which might have been used for an engine and no sign of any propellers of any kind, although* **at least one paper fin had been glued onto some of the tinfoil."**
- *"There were no words to be found anywhere on the instrument, although there were letters on some of the parts.* **Considerable scotch tape and some tape with flowers printed upon it had been used in the construction."**
- *"No strings or wire were to be found but* **there were some eyelets in the paper** *to indicate that some sort of attachment may have been used."*[30]

This is based on reporting that was done very recent to the event, which makes it far more reliable than the extraordinary recollections that would come many years later. Did you notice the absence of any talk about dead alien bodies in or around the wreckage? Don't you think something like that would have been prominent on the mind of Brazel and others? But they seem concerned only with rubber, string, and Scotch tape. Another key detail is that the men tried to *construct a kite* out of the materials. They failed, but their attempt says everything we need to know about what planet this wreckage likely came from. Clearly, they did not see this as the remains of a spacecraft. Just imagine if you found a crashed space shuttle in the desert. What are the odds that your first impulse would be to reconstruct a *kite* out of the debris? While I am sure that any species capable of traveling to Earth from another world would be significantly more advanced than we are, it's doubtful that they would be so smart as to be able to make it here in a vehicle constructed out of rubber and wood and held together by glue and tape.

Continuing with the story, the materials were then flown to Fort Worth Army Air Base in Texas, where people there immediately recognized what it was. They took one look at the paper, foil, rubber, and balsawood sticks and issued an official report declaring that it was nothing more than debris from a downed "weather balloon." That's it; case closed. Most people accepted that and forgot about it until the Roswell crash myth grew many years later. But there is more to the original story.

The military lied. Yes, there really was a Roswell cover-up. Today the whole truth is known, and it's clear that Brazel *did not* find the remains of a mere weather balloon. Sadly, however, the real story doesn't include aliens crashing on Earth. This probably explains why it's not nearly as popular as that other Roswell story.

The only thing the military was honest about was the "balloon" part of their explanation. What they left out is that this balloon was far more interesting and unusual than some run-of-the-mill weather balloon. The debris was almost certainly from one of the many balloon-supported listening devices used in Project Mogul. This top-secret program utilized many large balloons that rose to extremely high altitudes. In some configurations, multiple balloons were connected by chord to form vast "flight trains." Electronic listening devices were attached to a trailing cord for detecting the sounds of distant explosions, such as above-ground nuclear bomb tests in the Soviet Union. Remember, this was in the mid to late 1940s. Back then, there were no spy satellites or long-distance supersonic spy planes like the U-2 or the SR-71. But the Cold War was underway and the United States was determined to keep an eye, or ear in this case, on its rival. At this time, the United States was the lone nuclear power and was very concerned about the Soviets getting the bomb. All of this was extremely secret work, so it should surprise no one that the military lied.

B. D. Gildenberg, a participant in Project Mogul, says the work was so secret that even many of the people working on the program didn't know its name or completely understand what it was all about.[31] Project Mogul wasn't declassified until 1972. The reason for the extreme secrecy was that the United States didn't want the Soviets to know they were listening, because then they would have begun testing underground and made it much more difficult for the United States to monitor them. Gildenberg is certain that what Brazel found in 1947 was the wreckage from one of the kite-like radar reflectors that also hung below the balloons.[32] These reflectors were designed to enable ground crews to track the balloons with radar. They were constructed out of rubber, balsa-wood sticks, paper, foil, tape, and glue. Sound familiar?

A big clue that the elaborate Roswell UFO crash story is a made-up tale is the fact that *no one* in 1947 said anything about seeing an obvious spaceship or alien bodies being recovered and sent to Area 51 for study. All those juicy elements didn't emerge and become standard components of the story until *thirty years* later. Only after UFO belief grew, Hollywood unleashed its barrage of alien sci-fi films, alien abduction stories were told, and the classic big-head and tiny-body alien became a prominent fixture in pop culture did the Roswell story morph into what it is today.

Other secret military projects in the region after 1947 may have helped inflate the myth as well. Probably none did this more than the testing of high-altitude ejections systems for pilots. I interviewed former test pilot Joe Kittinger in 2001 about his remarkable aviation career, and he confessed to possibly having played a role in the Roswell myth. He said he and his team dropped short dummies, dressed in futuristic-looking flight suits, from high altitudes for various experiments and tests. Could local people have mistaken some of these dummies for aliens? "Absolutely they did," Kittinger said. "These dummies that we dropped from balloons were dressed in [silver] pressure suits, so they looked unusual. One time we dropped one and it fell way up in the mountains. These dummies weighed more than two hundred fifty pounds. So how do you carry one out of the mountains? We put it on a stretcher and carried it to the back of an ambulance to take away. Now if somebody is back in the weeds watching this they are going to say, 'Wow, look at that alien they have there.' We think that a lot of the alien sightings were actually us doing our work with the test dummies."[33]

This was several years after the original "crash" in 1947, however, so how could people remember seeing aliens/dummies in 1947? Because that's how human memory works. It is well established that the best of us can and do get the timelines of past events confused. It is very easy, natural you might say, to make the mistake of remembering that something happened twenty years ago when it actually happened only five or ten years ago. It is also very easy for the memory of a real event to be contaminated by fictional information obtained from a movie or a book. And it can feel absolutely accurate, no matter how terribly wrong it is.

The Roswell Incident enjoyed a burst of renewed media and public attention in 2011 when Annie Jacobson's book, *Area 51: An Uncensored History of America's Top Military Base*, was published. Much to the surprise of many readers, her mostly straightforward and credible history of the secret and important work that went on for decades there included a bizarre story about the Roswell crash near the end of the book. According to the claim that Jacobson presents in a manner that seems intended to convince readers, there really was a flying saucer and tiny humanoids really were recovered at the site. But they weren't from outer space. According to Jacobson's unnamed informant, the spaceship was an advanced flying vehicle built by the Soviet Union and sent into American airspace in order to test US defenses and/or terrorize the US military and public. The crew was not made up of aliens but of human children who had been surgically and/or genetically altered by Soviet scientists to look like big-headed, big-eyed extraterrestrials. Josef Mengele, the infamous doctor who conducted experiments on Jews at Auschwitz during World War II, was in Soviet hands then and helped transform the unfortunate kids into alien invaders.[34] Great story! What is not so great, however, is any suggestion that it should be accepted as reliable or true. An extraordinary story from an anonymous source—no matter how exciting the story may be—is not good evidence, proof, or anything else other than just another tall tale to be added to the billions of others that humans have been telling for millennia.

CONSPIRACY THEORIES

I was fascinated by the speed of the conspiracy theories that were generated after the Sandy Hook Elementary School shooting in 2012. Within days, if not hours, of the event, claims of lies and cover-ups were flying around the world. *An Israeli assassination squad did it to stir up international problems. The US government did it in order to have an excuse to impose radical new gun-control laws leading up to the confiscation of all privately owned guns.* Against all logic and without good evidence, conspiracy theorists said no children had been killed at the school. There were no grieving parents, no distraught teachers. The people seen on television and interviewed by journalists were "crisis actors" playing roles designed to dupe us. Sandy Hook—like the Kennedy assassination, the 9/11 attacks, and thousands of other terrible events—will forever have crude, disturbing, and unproven conspiracy theories attached to it. Why does this happen over and over? Why do people believe these kinds of claims?

Before analyzing the conspiracy-theory phenomenon, it is important to be clear that evil, destructive, and criminal conspiracies are very real. They happen all the time. Of course groups of people get together to plan and execute bad deeds. We are social creatures—for better and for worse. Both history books and today's headlines offer countless examples of mischief by committee. Most popular conspiracy theories are not like supernatural/paranormal claims because they don't seem to contradict what we know about how the natural world works. The problem is that many of us seem to have an irresistible urge or need to have complex and dramatic explanations when bad things happen, even if they lack proof and on the surface appear to be illogical and very unlikely. This should surprise no one. The following *standard* human characteristics make conspiracy theories of interest to most and irresistible to some:

- **We love stories.** Telling stories and listening to stories is a fundamental aspect of being human. We are storytellers. Gossip isn't just fodder for the tabloids and simple minds. Gossip is a crucial glue that binds us to one another. Since deep into prehistory, stories and gossip have been the devices by which we have shared important information, alerted each other of dangers, and inspired one another. One thing all good conspiracy theories have in common is that they are *great stories*. There usually are victims, villains, secrets, and a chance for justice to prevail—if only enough people will believe the story.

- **We don't like little answers to big questions.** It insults our sense of justice to accept that Lee Harvey Oswald, just one man, was able to kill the most powerful man in the world. There must have been many powerful people involved, right? Not necessarily, because human history and contemporary culture are not math equations that must balance in a way that conforms to our subjective tastes. Sometimes unlikely things just happen, and sometimes events violate our perceptions of what is fair and reasonable.

- **Those damn dots again!** Not only can we see things that aren't really there by connecting "dots" that don't really have a connection, we often *think* that way, too. Our pattern-seeking ways do not end when we stop looking at clouds and inkblots. Sometimes we connect random dots within stories and information, too. With our minds, we seek out links between facts, comments, events, and people. If enough forced connections are made, then a conspiracy may come into focus, whether or not it's really there. Just like our brains create and see the shape of a monster in a passing cloud, many of us create and see a monstrous secret in big events, even though it's not real.

Because conspiracies really do happen, it is important for people, skeptics included, to avoid lumping all conspiracy theories together. Many of them deserve a hearing, not only for the sake of fairness but also to help counter claims of a cover-up. I think it's crucial to continually steer away from emotion and conjecture with these ideas and stay focused on hard evidence. It is amazing what people will think and say when it comes to conspiracy theories. In 2012, I was a guest on the AM radio show *Coast to Coast* to discuss skepticism and science.

We talked about many topics, but the conspiracy stuff generated the weirdest comments in the days and weeks after the show. I received e-mails and saw web posts that accused me of being a paid stooge of "Big Pharma," out to promote profitable vaccines and spread disinformation about alternative medicine. My view that al Qaeda was probably responsible for the 9/11 attacks and that the Moon-landing-hoax claim is almost certainly wrong led some to declare that I was "obviously" a government-paid "black operator" on a mission to mislead the public. Clearly some believers have allowed emotion and suspicions to distract them from the need to focus on evidence. However, some conspiracy theorists don't have that problem. They go to extremes with evidence. They have mountains of evidence and will happily bury you with it if you allow them to. In my experience, the proponents of conspiracy theories who do the hard work of accumulating evidence are far from dim. They tend to be bright, energetic thinkers. What they seem not to realize, however, is that conspiracy theories are the near-perfect trap for confirmation-bias problems. When one analyzes a complex crime, large organization, or major event, it is all too easy to cherry-pick the data in order to present a lopsided case. I'm not suggesting that they are dishonest. It is only natural to subconsciously do this. Confirmation bias is universal. Conspiracy theorists, no matter how intelligent and how sincere, certainly are not immune to it.

The best advice I can offer about conspiracy theories is to keep an open mind, because groups of people, corporations, religious organizations, and governments really are capable of doing just about anything. History proves it. But don't embrace a wild conspiracy claim because you think it *could* be true or because you feel it *should* be true. Play hard to get. Hold out for proof.

ASTROLOGY

Astrology is the claim that bodies out in space influence bodies down here on Earth. This amazing idea has existed for thousands of years. I get conflicting explanations today, however, when I ask believers if these celestial bodies cause our actions or simply foretell what is going to happen without actively causing any of it. Either way, it's fascinating to imagine such a spooky connection with stars and planets. There were times in our past when astrology made sense to even the most educated and enlightened people in many societies. It was viewed as respectable science and a productive means of gathering information. But we have learned a lot about both space and human psychology since the days of ancient Babylon. For example, scientists figured out some time ago that the location of stars and planets show no detectable or measurable impact on the personality and fates of individuals. Really, it's true; Mars and Saturn have nothing to do with whether or not you will get that promotion at work or marry the ideal spouse one day. Venus has no bearing on how high-school cheerleading tryouts will turn out for you.

Astrology presents us with yet another opportunity to pause and reflect on the profound weirdness of humankind. On one hand, we have come so far. People once looked up at the night sky saw gods and animals in the stars. They imagined that they cared about us, influenced us, and whispered secrets. Later, we would walk on the Moon, land robots on other worlds, peer deep into the universe and time itself with powerful telescopes, and even contemplate the end of this universe. Yet for all the brilliance we have shown, many millions of us still look up and see planets and stars that know the intimate details of personal lives.

It is important to be absolutely clear about one thing: Astrology is not scientific. It is not evidence based. It has never been proven that astrology is able to describe our lives or predict the future, so I don't think it can. But don't take my word for it. Make up your own mind.

Beyond the fact that it's unproven, there are two key points that crush the case for astrology. First, its origin reveals just about everything we need to know about how valid it is likely to be. Astrology's earliest beginnings are lost forever in prehistory. But it's not difficult to imagine early humans injecting personal meaning into the visible stars, planets, and the Moon. Who can blame them? They knew nothing of astronomy, so the beauty and mystery painted across a typical night sky could easily have sparked magical thinking. *These lights above me must mean something. How could they not? Perhaps I am connected to them. Maybe I should fear them, revere them, or use them to understand the meaning of my life. They must be the key to understanding this confusing world I live in. They are telling me something.*

History's earliest mentions of astrology come from the Sumerian and Babylonian cultures, stretching back three thousand to four thousand years ago. Astrology in one form or another has been hugely influential throughout history and is in some ways the ancestor of astronomy. Try not to read too much into that, however. It's like saying that primate grooming—the practice of picking and eating lice and ticks off your buddy's hide—is the ancestor of haircuts and shaving. True, perhaps, but not a pastime we necessarily need to hold onto today. In most societies and for many centuries, everyone from queens and kings to generals and doctors did little without consulting an astrologer. Even in modern times, some of the world most powerful leaders have been influenced by astrological beliefs.[35] But don't allow yourself to be overly impressed by age and influence. Popularity and some degree of historical impact are not enough to save astrology.

The ancient Babylonians never conducted scientific studies that linked personality types to the movements of celestial bodies. The Sumerians did not amass data that correlated past historical events with star positions in a way that could be used to reliably predict future historical events. No, they just looked up at the stars, played connect-the-dots, and declared that there was meaning in what they saw or imagined they saw. If a modern-day astrologer claims to rely on some book or source documents, ask for titles. Ask where it or they came from. Ask for early sources that do not simply describe astrology but also explain its origins and how it functions in a way that makes sense. Follow the trail back far enough, however, and it always leads to a familiar place: human imagination. Astrology is layer upon layer of *made-up stuff*. This is not science. Its predictive powers do not exist. Astrology has nothing to do with the stars and planets of outer space. It has everything to do with the biases and delusions of human *innerspace*.

Astrology's second fundamental problem is that nobody seems to know how it is supposed to work. "It just does," is the most common explanation I hear from believers. Astrology may have been important to the development of early astronomy, and it may have even been useful in keeping time and developing early agriculture. However, as it is practiced and believed in today, it's pure pseudoscience. There is no underlying theory that explains it, no body of knowledge that supports it, and no set of experiments that confirm it. I know this because I have asked professional astrologers. I listened with an open mind, and when told that my status as a Libra meant that one thing or another would happen to me and that I was inclined to think and feel this way or that, I asked how they know this. "Because the stars say so," is not an answer, of course. *Why* do the stars say this? *How* do the stars say this? These are the simple questions that need to be answered, but never are. I suspect that many believers imagine there must be some vast collection of data stored in the vaults of a giant astrology archive somewhere that serves as the source material for all the horoscopes written today. Or maybe they think there is some centralized supreme council of astrologers that generates the calculations all astrologers base their work on. But there are none of these things.

A final point that everyone should be aware of is that horoscopes are popular and work well for so many people because most of them are written in ways that allow virtually everyone and anyone to recognize something that feels unique and special to them. A good horoscope writer can craft fifty words or so that will feel personal and specific to the individual reading or hearing it—no matter who that individual is. If something is known about the person in advance and the horoscope can be personalized, even better. It's like psychics with their cold readings. Focus on challenges and concerns common to most of us, and those who are predisposed to believe will think it must apply to them in some special way. As far as predicting things about the future, two things help that along: vague predictions and confirmation bias. A prediction like: "You will enjoy financial gain this week," could mean winning the lottery, finding $5 on a sidewalk, or nothing more than getting your weekly paycheck. For most believers, confirmation bias will ensure that astrology's relatively few successful predictions are remembered and its many failures forgotten.

When someone tells me how awesome astrology is, I don't argue. I just ask a few questions and pay close attention to the answers. Try it. Ask how the stars and planets go about influencing or predicting the course of our lives. What is the force behind this? How does it work? Is it gravity? If so, how can that be when a potted plant in the room exerts more gravitational force on you than a distant planet? Ask them why it is that people born in October have different personalities and talents than people born in December. Don't let them ramble on about all the *ways* in which they are different; you want to know *why* they are different. Asking these important questions and not allowing yourself to be sidetracked by hollow answers are key to becoming a good skeptic and staying focused on reality.

MIRACLES

Miracle belief is extremely popular. Just look around, just listen. Miracles are everywhere. One can hardly walk down the street or turn on a television without hearing someone claim that a miracle has occurred. According to Harris Poll, 76 percent of American adults believe in miracles.[36] But believing in miracles is easy. *Thinking* about miracles takes a bit of work, and, despite all the talk, most people don't bother. They just seem to accept them as a fact of life without question. But a good skeptic doesn't believe in something only because most people around her say it's true.

Miracles are generally thought to be supernatural events caused by a god, a magical being, or a person with special, unnatural powers. According to various believers, miracles can be bizarre feats that defy the normal workings of nature. They can also be natural events that are intentionally orchestrated to bring about desired outcomes. So a miracle might be a natural rainstorm that a pious farmer's prayers bring on just in time to save his crop or it could be a god or preacher bringing a dead man with rigor mortis back to life by supernatural means.[37]

Maybe miracles really do happen. I don't know. But I am sure that the claims of miracles that I hear about are obvious cases of people failing to think critically and be good skeptics. Just because something is unexpected, unusual, or extraordinarily well timed does not necessarily mean that it is a magical event. Of course weird things happen all the time. Wouldn't it be strange if nothing strange ever happened?

A key problem people have when thinking about miracles comes from the fact that we aren't natural mathematicians. Most people in the world never take a class in statistics or even pause to consider how likely it is for "unlikely" events to occur. Just consider, for example, that there are seven billion people alive right now. Let's imagine some quirky event that happens every day but at a very low rate—let's say this event has a *one-in-a-billion chance* of happening to you or me on any given day. It probably won't happen to us, but will happen to *seven people* every day. Those seven people are probably going to feel extremely lucky or unlucky, depending on what it is. But it was going to happen to somebody. Even a rare daily event with one-in-a-million odds of happening to you or me sounds unlikely, right? But *seven thousand people* would experience the event each day. Over the course of just one year, 2,555,000 people would experience it! It's the same with lotteries.

The chances of winning those things tend to be astronomically small. But people do win them virtually every day, nonetheless. No doubt, the winners feel very special and many of them likely conclude that they must have been on the receiving end of a miracle. But when you consider those few winners in the context of all the losers, it becomes clear that nothing magical necessarily occurred. *Somebody* was going to win because lotteries are designed to produce winners.

If you are prone to believing in miracles and you think some things happen that just can't be explained as anything other than the work of a god, then try to imagine being in very different circumstances. Transport yourself into the body of a typical Pleistocene human one hundred centuries ago. You are living in a time before Google, *Wikipedia*, and soap. Your clothing, footwear, and scent will change, but biologically you are no different. Mentally and physically, you are the same. What you are missing out on, however, is the convenience of being able to call on 100,000 years of accumulated human knowledge and scientific discovery. It's gone. All you have now is the potential to reason for yourself and figure things out. You also have whatever skills and awareness your prehistoric peers passed on to you around nightly campfire gatherings.

Suddenly a nearby volcano erupts. The ground shakes. A strange, orange-and-black goo flows down the mountain. When it comes into contact with trees, they burst into fire. Smoke and dust darken the sky. You are terrified but also desperately curious. *What is going on? How can this be happening? It makes no sense!* All you have are questions and no answers. You are smart, but you know nothing of geology, volcanology, the layers of the Earth, magma, lava, and so on. You can't simply ask someone else to explain what is going on because no one on the entire planet knows. What do you do? You might have a flash of insight and come up with a respectable guess about the Earth's fiery underbelly releasing a bit of pressure. But even then you couldn't be sure. No one can look at an erupting volcano for the first time and explain it sensibly and fully with absolutely no knowledge of geology to draw from. So maybe out of frustration you would do what everyone around you is doing: explain the volcano as the work of gods or magic. "The gods did it" and "It's a miracle," has probably been the most common quick answer to every mystery faced by people since the dawn of language. But from the perspective of your presence in the twenty-first century, you know that a volcano can be explained by natural forces and doesn't require the involvement of gods or magical forces. Nothing would have changed about you from the Pleistocene and now other than the *availability of a natural explanation*. This is key because as natural explanations come, supernatural explanations go. This has been the pattern for thousands of years. Magical explanations lose their appeal when they just aren't necessary anymore. Thousands of years ago, who wouldn't have thought that a solar eclipse was a miracle? Today, however, fewer people in most societies feel the need to invoke magic as an explanation for them. Nothing changed in nature. Only our knowledge of astronomy did. Perhaps patience is the first thing we should think of when faced with unanswered questions. If the past is a guide, many of today's miracles probably won't seem so miraculous in the future.

It should not be interpreted as insulting or rude to point out that ignorance drives many specific miracle claims and the general belief in them. Ignorance doesn't mean stupidity or intellectual dimness. It only refers to a lack of specific knowledge needed to understand something in a given situation. We are all ignorant of many, many things, of course. What matters is how we choose to react to our ignorance when it stares us in the face. Some of us freely admit that we do not know some things and proceed accordingly. Others, however, attempt to hide their ignorance under a tent of made-up answers.

It seems to be human nature, or an irresistible impulse, for us to explain things even when we can't. So we make up answers. We fill in the blanks with "miracle." But here's a critical truth every good skeptic knows: It's okay to say "I don't know." Really, it's fine. I understand that it can be difficult at times, but this is how humans ought to operate. If I were out walking my dog, and an autographed first-edition copy of Mark Twain's *The Innocents Abroad* suddenly fell from the sky and landed at my feet, I would be challenged to come up with a natural explanation. *It fell out of a passing airplane currently being leased by a wealthy collector of rare books? It was swept up into the stratosphere by a tornado in Oklahoma and then it blew over to my state before tumbling earthward? Some juvenile delinquent stole it from the Twain museum in Hannibal, Missouri, and then traveled to my neighborhood so that he could throw it at me while perched in a nearby tree?* All sound and reasonable possibilities, of course, but none of those would probably be the right answer. Maybe I would just have to live the rest of my life with a nagging mystery, one that gnaws at me every time I hear the name Mark Twain or look at that book on my shelf. As a good skeptic, however, I am sure that I wouldn't cheat by pretending to know that the gods tossed it down to me as gift. I wouldn't do that because I couldn't possibly *know* such a thing. Mysteries are not miracles.

ANCIENT ALIEN ASTRONAUTS

Did space travelers from a distant world visit the Earth during ancient or prehistoric times? Did they interact with the people, influence their engineering, art, and religious beliefs, and perhaps even interbreed with them to supercharge our evolution?

Maybe. Who knows? I can't say for sure that it didn't happen. What I do know is that no one has ever proved this amazing story or even presented any good evidence for it.

Chariots of the Gods?, a 1968 book written by a former hotel manager named Erich von Däniken, is more responsible than anything or anyone else for making this claim so popular. It was an international bestseller when it first came out decades ago, and it is still selling today. But why? Surely people would not just accept a claim this wild without rock-solid proof, right? No, they would and they do. Apparently it doesn't matter that virtually every professional archaeologist in the world rejects this claim. Nor does it seem to make any difference that every single argument Von Däniken made has been thoroughly addressed and demolished by experts. People seem to like this idea of ancient aliens visiting us so much that they can't resist boarding the chariot.

I won't go into detail about the numerous problems with the ancient-astronaut claim. Those who want more can read the chapter about it in my book *50 Popular Beliefs That People Think Are True*. The point I want to make here is that this story is rooted in a view of our ancestors that is both inaccurate and detestable. Von Däniken and others who promote this belief push the premise that prehistoric and ancient people were hopelessly dim and could never have done the things anthropologists and historians say they did. These were such bumbling idiots that they had to have had extraterrestrial help in order to get anything done. If not for the advanced visitors, it's clear that we would all still be grunting at one another in dark caves or up in trees. Here is a small sample of claims made about ancient people found in *Chariots of the Gods?*

- "In the Subis Mountains on the west coast of Borneo, a network of caves was found that had been hollowed out on a cathedral-like scale. Among these colossal finds there are fabrics of such fineness and delicacy that with the best will in the world one cannot imagine savages making them."[38]
- "The Great Pyramid is [a] visible testimony of a technique that has never been understood. Today, in the twentieth century, no architect could build a copy of the pyramid of Cheops, even if the technical resources of every continent were at his disposal."[39]
- "Where did the narrators of *The Thousand and One Nights* get their staggering wealth of ideas? How did anyone come to describe a lamp from which a magician spoke when its owner wished?"[40]
- "It is an embarrassing story; in advanced cultures of the past we find buildings that we cannot copy today with the most modern technical means."[41]

Not only are these assertions condescending, but also they're just plain wrong. First of all, anatomically modern people have been around for at least 200,000 years or so. This means that people with brains as capable and creative as ours were here for more than 195,000 years *before* the great pyramids of Giza were built and stories like *The Thousand and One Nights* were written. Interestingly, prehistoric people had slightly *bigger* brains than we do today, and they might have been more intelligent.[42] This makes sense when you consider how unforgiving life would have been for Stone Age dimwits. Most would have been selected out of the gene pool early by poisonous berries, falling rocks, and hungry predators. Today, however, the dimmest not only survive at high rates, but some of them belong to an extremely successful subspecies known as "celebrities" and star in reality TV shows.

No one with any sense has ever suggested that we went from Australopithecines sleeping in trees to *Homo sapiens* building pyramids in one flash of insight. A lot happened between *Homo erectus* and the reign of Khufu. The progression of technology was not so unusual that it requires extraterrestrials to explain it. In short, there is nothing about ourselves or our past that indicates that we were incapable of doing the things we did. And, contrary to what Von Däniken says and writes, we could in the past and can today build a giant pyramid if we wanted to. Researchers have even demonstrated ways in which people might have moved and placed large stone blocks *using only ancient technology and muscle.*[43] There is no doubt it could be done with modern technology.

Even within Egypt, the progression is evident. The three famous pyramids at Giza did not spring up from the sands in a way that defied the flow of ancient Egyptian history. I have visited the nearby Pyramid of Djoser at Saqqara, for example, and this older pyramid is precisely what one would expect to see. It's huge and impressive yet less of a technical and artistic achievement compared with the pyramids that followed. What are we to believe—that less intelligent and less capable aliens helped the Egyptians at Saqqara earlier and then, years later, slightly more sophisticated aliens lent a hand at Giza to produce the superior pyramids?

THE END OF THE WORLD

Writing about science can open one up to a universe of opportunities. For example, I regularly get invites to talk about everything from astronomy to zombies.[44] In 2012, I was one of the speakers at *Doomsday Live!* a radio show and live-audience theater event organized by the SETI (Search for Extraterrestrial Intelligence) Institute.[45] It was a slick production, hosted by the Computer Science Museum in Mountain View, California. The topics seemed thrilling enough to me: asteroid strikes, viral plagues, rebellious computers, environmental chaos, irrational doomsday predictions, and more. But it might easily have been a disaster about disasters. I wasn't sure how people would react. Fortunately, it turned out to be one great apocalyptic spectacle. The live audience was one of the warmest and most enthusiastic I have ever spoken to. At times, I felt like I was at a Super Bowl bash or maybe one of those Tony Robbins seminars. Yeah, there is just something about the end of the world that picks up a party. But I should have expected it because this is nothing new.

Apocalyptic fears and fantasies have intrigued, excited, burdened, and haunted humankind for millennia. I would not be surprised if *Homo erectus* bands often paused between hunting and gathering to worry about the end of the world, too. Maybe caves were the first doomsday bunkers and cavemen the first doomsday preppers. What is certain is that we have a strong and enduring interest in a final day of reckoning that borders on obsession. Just note the steady flow of films, books, and television shows with apocalyptic themes in pop culture today. Numerous religions, past and present, were built on a foundation of end-of-the-world threats and promises of escape. Science has done its part too by providing several charming, evidence-based doomsday scenarios for us to contemplate at bedtime. I understand the appeal. Horrible as global death and destruction may be, there is also a certain attraction to it for some people who think the idea of hitting the reset button might not be so bad. No more rent or mortgage payments to worry about. No lame job to show up for anymore.

The end of the world, human extinction, and the collapse of civilization are all interesting subjects for the skeptic because the world really could "end" and take us with it. It's not an entirely crazy topic. For example, gamma rays, a nearby supernova, and solar flares are no joke. They are all real things that do happen. One event could scorch us into oblivion. Some threats are not only possible or likely to happen; they *will* happen. Wandering rocks in space called asteroids will threaten to devastate our planet and destroy most life here one day, just as they have done in the past. Right now, we know there are approximately a million objects in our solar system that could pose a direct threat to us one day. We might be in the crosshairs in a few years, a few decades, or a few million years from now. But it's going to happen.

Supervolcanoes are *thousands of times* more powerful than the regular volcanoes we are familiar with. These gigantic monsters are rare, but they have erupted in the past and will erupt again. One of them, Toba in Sumatra, almost wiped us out 74,000 years ago when it spewed tons of dust and smoke into the atmosphere to darken and chill the world. The total human population may have dipped to as few as two thousand people worldwide as a result. It is unlikely but plausible that a virus could evolve or be engineered to kill us all one day. Nobody seems to worry about nuclear war since the Cold War ended, but it remains probably the greatest threat to civilization right now. It's mostly out of the public's thoughts these days, but the United States and Russia still have thousands of nuclear weapons ready to fly—by order or by accident.

My favorite end-time scenario, by far, is human extinction via a massive "methane burp" from the seafloor. Yes, believe it or not, the gaseous residue of hundreds of millions of years of microbial activity beneath the bottom of the ocean might be unleashed in one colossal fart that kills us all by abruptly altering the atmosphere and climate we depend on. Critics less forgiving than me might feel it would be poetic justice, a fitting end for the species responsible for two world wars and too many boy bands.

Don't panic and run screaming into the streets just yet, however. Even these frightening, evidence-based scenarios probably wouldn't cause our extinction because we may develop the means to predict, avoid, eliminate, or lessen their impact. This is why it is vital for us to continue exploring and discovering. Some of the answers we find today may save us tomorrow. There is also strength in numbers. Remember, there are more than seven billion of us now, and we are spread across much of the planet. Our large population, combined with our creative intelligence, makes us a very tough species to kill off. A 2009 University of Colorado study supports my view on this, concluding that we are "unlikely to become extinct without a combination of difficult, severe, and catastrophic events."[46] The researchers said that it was hard to come up with plausible scenarios capable of killing *everybody*. But this doesn't mean that the world isn't doomed. In fact, it almost surely is.

In four or five billion years, our star will die like all stars eventually do. But the Sun won't go quietly. It will behave like an angry man dying, punching and flailing at anyone near. After all the life-giving support it provided for billions of years, the Sun's final act will include frying the Earth and boiling away all its water. But there is still plenty of room for hope. Four billion years is a long time. It is unlikely, if not impossible, that we would be close to anything resembling what we think of as human today, if we still exist. Maybe by then our decedents will be living comfortably on a safer planet far away or on an artificial world they built. Maybe they will have become so technologically advanced that they can control the Sun or somehow protect the Earth from its fury. But even if we can dodge the Sun's wrath, we might not be able to escape ultimate doomsday. Current thinking among astronomers says our expanding universe is likely to go dark in about 100 trillion years or so. Still we might escape by jumping to another universe or circling back in time. Then again, maybe humanity—along with all our accomplishments, creations, and memories—will be erased from the cosmos forever. Time will tell.

Evidence-based predictions of possible doomsday scenarios, while fascinating and well worth concern where appropriate, just are not a big problem for very many people. If one thinks critically and carefully about such ideas, and assesses the danger like a good scientist might, then there's not much to lose sleep over. Yes, we need to improve our early-detection abilities for near-Earth objects that may threaten us. We also need to be ready to respond effectively when they do. Yes, we should do more to discourage and prevent people and nations from weaponizing viruses and bacteria. We also might want to reduce nuclear arsenals to only enough firepower needed to destroy civilization, say, once or twice and no more.

The key to dealing with this subject sensibly is, of course, science and skepticism. When irrational thinking creeps in, apocalyptic thoughts not only become silly, but also can turn dangerous. There are many examples of irrational believers doing terrible things in the name of doomsday. To my knowledge, however, there are no suicidal groups based solely on a scientific reading of the Sun's expiration date. There are no apocalyptic religions centered on what geologists say about supervolcanoes. I doubt anyone has quit his job and sold his house in anticipation of the big methane burp or the rise of artificial intelligence. No, the *science* of doomsday, while important, is just not a popular source of self-destructive behavior. It is belief in *unscientific* doomsday claims that consistently causes the trouble.

Far back in time, during the first days of civilization or even earlier perhaps, some man or woman told anyone who would listen that the world was about to end in some spectacular way that would kill everybody. And people bought it. So began a weird and steady freak show of human behavior. For generation after generation, in society after society, hundreds of millions of people have believed in countless doomsday claims. They always come with excessive drama and are delivered with unwavering confidence. But good evidence and logical arguments are never part of the presentation, which probably explains why the failure rate of these predictions is 100 percent to date. Why does this diseased thinking keep infecting us? I'm not sure, but I do know that skepticism is the cure.

It's impossible to know accurate numbers, but there is no doubt that throughout history many millions of people have suffered and/or died as a direct result of belief in irrational, never-going-to-happen doomsday declarations. Does our fascination with death somehow derail our ability to recognize and reject stories that are clearly made up? Maybe all or most of us come with a built-in doomsday fetish that makes this weird seduction inevitable. Maybe our fondness for patterns leads us to imagine a near and specific end to the human story, just for the sake of completing the picture or closing the loop.

The good news is that you never have to fall for these stories. All you need to do is think like a scientist, and the madness bounces right off. Skeptics are the last people on Earth who will burn up nervous energy, throw away money, join a kooky organization, kill themselves, or harm anyone else in anticipation of a doomsday prediction that no one is able to back up with good evidence. We know that it makes sense to assume that there will be a tomorrow. Skeptical thinking has defied doomsday and saved the world many times before and can again. All we have to do is think, and those scary predictions wither and die.

MOON-LANDING HOAX

How did this happen? How could humankind's greatest adventure and greatest technological achievement end up being called a hoax and a secret conspiracy by millions of people around the world? The Apollo program landed men and on the Moon six times! The astronauts talked to us from the Moon. They took photographs. They filmed their moonwalks. They brought back nearly a thousand pounds of rocks. Nevertheless, many people deny that the Moon landings ever happened. They charge that it was an elaborate trick designed to make the United

States look good during the Cold War. President Richard Nixon and NASA lied to us. The astronauts were actors.

As we covered in the section on conspiracy theories, great stories are irresistible to the human brain. We can't help ourselves. We have to listen to stories and gossip. We can't stop ourselves from considering them and imagining if the wild claims are true. The juicier, the better. The urge to believe in and spread gossip is probably instinctual, just part of who we are, say some researchers.[47] It's less about what we do and more about who we are. "Gossiping Primates" is not only a great name for a rock band, it's also who we all are deep down. In light of this, I think that something like the Moon-hoax claim should be viewed as less a kooky aberration and more an inevitability. The Apollo missions were too big, too complex, too difficult, and too spectacular for everyone to accept. When the moment arrives that a human finally walks on Mars, I guarantee that some subset of the world will say it's happening on a Hollywood soundstage.

Because I am committed to being an open-minded skeptic, I have always made the effort to listen to and consider the arguments of Moon-hoax proponents. No matter how ridiculous some of the ideas may seem to me, I don't automatically dismiss them without at least thinking them over. One never knows what will happen tomorrow, but this claim does not appear to be doing well in America today. Only about 6 percent of the adult population believe it was all a hoax, according to Gallup.[48] Although that figure represents millions of Americans, it's definitely a fringe number for a poll. One can ask just about anything and get at least 6 percent to agree to it. However, we also have to consider how many people doubt the landings, if not deny them. There is also the question of how many non-Americans don't believe anyone has been to the Moon. I have lived outside the United States and traveled extensively, and many times I have been shocked to discover how widespread this belief is. Based on my experiences, it's very common in many Asian, African, and Caribbean countries. One poll found that a quarter of all British people are Moon-hoax believers.[49] Motivations for this belief vary, I'm sure. It might be ignorance about science (*What's the Moon?*), poor knowledge of history (*President Nixon was impeached for his role in the Moon hoax, right?*), lack of imagination (*There's just no way!*), or even anti-American sentiments (*I hate Americans, therefore they never went to the Moon.*). Late Venezuelan president Hugo Chavez was a Moon-hoax believer, probably in large part because he didn't like the United States.[50] Some people, members of the Taliban in Afghanistan are a good example, are so lost in medieval thinking and superstition that they likely view it as impossible for mere mortals to achieve such a voyage. For some it may be a result of the formal education they received. I had heard, for example, that the Moon-hoax claim is taught to children as fact in Cuban schools. I was very curious about this when I visited that country. But as an American citizen allowed in as a journalist, I wasn't in a position to stir up political debates, so I skirted around the issue. The Cuban people I encountered were consistently bright and had positive views about science—but some were unconvinced that NASA ever sent men to the Moon.

Young people today in all countries seem particularly vulnerable to this belief, perhaps in part because the landings happened in the 1960s and 1970s and therefore feel like ancient history to them, which makes it more difficult to believe. After all, they don't see anyone going to the Moon today. So how did people do it with "prehistoric" twentieth-century technology? *No cell phones or laptops, but they went to the Moon? Sure.* It doesn't help that any curious kid with a computer can watch or read Moon-hoax propaganda on the web. Without some honest context and sensible counterarguments, it can seem credible, I'm sure.

In 2001, Fox TV broadcast a terrible pseudo-documentary during primetime called *Conspiracy Theory: Did We Land on the Moon?* It probably helped inspire a new wave of irrational deniers then, and it's still derailing brains online today. YouTube clips from it have drawn hundreds of thousands of views. Professional presentations work, regardless of how factually incorrect or dishonest they may be. One survey found that 27 percent of young Americans aged eighteen to twenty-four had "some doubt" that astronauts went to the Moon, and 10 percent said it was "highly unlikely."[51]

When I read about such statistics or meet people who believe in the hoax, my first reactions are sympathy and disappointment. I feel sad for those who feel no connection to this important accomplishment. It was a human feat that all people should at the very least take some measure of pride in. Sure, it was motivated by Cold War militarism and shallow nationalism, but it was still profoundly special and meaningful. The moment the first humans landed on the Moon, we were no longer an Earth-bound species. We had extended our physical reach and in doing so expanded our collective ability to dream bigger dreams. Or at least that's what should have happened.

The following are a few of the more common Moon-hoax ideas that I have heard, followed by brief responses. Remember, as with everything else in this chapter, I'm not encouraging you to agree with me "because I'm right." Whether you're in your sixties or in the sixth grade, I want you to think for yourself and make up your own mind. So when I share personal insights and experiences, please don't think that I am suggesting that you follow my lead because I'm some kind of an authority figure. You need to make up your own mind about this and every other extraordinary claim.

Where are the stars? This is probably the most common of all the challenges to the Moon landings. It's also the easiest to explain. If those astronauts were really on the Moon, hoax believers ask, then why are there no stars in the black space behind them? It looks suspiciously like the black wall of a movie-studio soundstage. The reason for this is simple. Stars in space are relatively faint dots of light. The surface of the Moon, however, is very bright when the Sun is shining on it. So too are the lunar module (landing vehicle) and the white spacesuits the astronauts wore on the Moon. Therefore, in order to take a decent photograph in such conditions, the camera's shutter speed must be set to a speed that is too fast to take in the relatively faint light from stars. The result is a photograph with properly lit astronauts but only black space behind them. If you still have doubts, try it yourself. Go out on a starry night and take a photo of a well-lit building or person. If the shot comes out well for the primary subject, you won't see stars. It's about differences in light. The contrast was so great, in fact, that Apollo 16 astronaut Charlie Duke says he couldn't even see stars with his eyes while on the surface of the Moon.[52]

The flag waved. There is a film clip hoax believers say shows the American flag waving in the wind on the Moon. Of course, this could not have happened because the Moon has no atmosphere, no wind. But there it is, plain as day for anyone to see, flapping around. Clearly somebody opened a door in the studio or turned on a fan while the fake flag planting on the Moon was being filmed.

Don't confuse moving with flapping in the wind. Yes, the flag moves. But why? It moves because an astronaut wiggled the flagpole while driving it into the lunar surface. There may not be wind, but the laws of physics are still enforced up there just like they are down here on Earth. The flag moves because of inertia. Movement of the flagpole results in movement of the attached flag. Simple as that.

Tricky Dick. Some people point to the man who was in the White House during the Apollo landings as a key reason to believe this conspiracy theory. But just because someone is known for lying and being involved in conspiracies doesn't mean *everything* he or she was ever involved with is necessarily corrupt. President Richard Nixon may have been an unethical politician, but this alone is not proof that he orchestrated six fake Moon landings. Watergate does not prove Moongate.

They lied. I have met many people who participated directly in America's space program during the Apollo years. I spoke with some casually at science conferences, others I conducted lengthy, formal interviews with for various writing projects. The following is a partial list:

Gene Cernan (Apollo 10, Apollo 17 commander)
John Young (Apollo 10, Apollo 16 commander)
Charlie Duke (Apollo 16)
Gene Kranz (Mission Control Center flight director)
Alan Bean (Apollo 12)
Buzz Aldrin (Apollo 11)
Scott Carpenter (Mercury Seven astronaut)
Tom Stafford (Apollo 10 commander)
Frank Borman (Apollo 8 commander)
Rusty Sweickart (Apollo 9)
Jim McDivitt (Apollo 9)
Walter Jacobi (engineer, member of Wernher von Braun's rocket
    team)
Dave Scott (Apollo 9, Apollo 15)
Ted Saseen (Apollo spacecraft engineer)
Jack Cherne (lunar module engineer)
James O'Kane (Apollo spacesuit engineer)

I think it is important to bring this up because if the landings were a hoax, it means that all of these people, and many more, are liars. If no one ever went to the Moon, then, for all these years, they have kept the greatest secret in history. Perhaps it's possible that they could be lying, but it does not seem plausible to me, because I looked in the eyes of astronauts John Young, Gene Cernan, and others as they shared personal memories of their time on the Moon with me. Prominent Apollo engineers told me about the hard work and long hours it took to create machines capable of carrying people to the Moon and returning them safely to Earth. Mission Control flight director Gene Kranz described to me how he *felt* during those long missions. It is easy to accuse men of lying when they are nothing more than printed names or two-dimensional faces on television. But when you meet them and listen to their heartfelt stories firsthand, it is difficult to imagine how they could all be liars. For example, Apollo 16 astronaut Charlie Duke told me that he left a photograph of his wife and children on the Moon. Seems like overkill if he's just protecting a hoax. It can't be for the money, because an Apollo astronaut could make millions by coming clean and exposing the conspiracy. Besides, why would they go so far above and beyond the call of duty to sell a scam after so many years have passed? The Cold War ended a long time ago. The Soviet Union doesn't even exist anymore. Kennedy, Johnson, and Nixon are all dead. I suppose it is possible that they all lied to me as they have to the world. Perhaps all of them are great actors. And maybe every one of them found the will to keep this secret for more than forty years. But I doubt it.

Where is the crater? If you look at photographs of the lunar module (the Apollo landing vehicle) on the Moon, you will notice that there is not a deep crater beneath it. *Why not?* ask hoax believers. Why didn't the lunar module, with its powerful rocket engine, blast a big crater in the lunar surface? The answer is simple: Those clever NASA engineers thought to install a throttle in the lunar module.

What happens when sprinters cross the finish line in a 100-meter race in outdoor stadiums? They slow down, right? They decrease their effort and coast a bit before stopping short of running into a wall or a fence. They certainly don't continue running at full speed until a wall or fence stops them. Same with the lunar module. When Apollo 17 commander Gene Cernan wanted to descend to the lunar surface, he throttled back for a gentle landing. He didn't make a crater because he didn't land with the engine at full output. He couldn't have, because an engine at full throttle would have made landing impossible.

As with many such odd beliefs, a little knowledge can go a long way with the Moon-hoax claim. Unfortunately, most Americans know very little about the Apollo program and know even less about the efforts of the US military and NASA that preceded it. They are also mostly unfamiliar with the Soviet Union's important and successful space program. The Soviets were very sophisticated and ahead of the United States during much of the 1960s. They almost certainly would have figured it out if the United States faked the landings, and we can be sure that they would have told the world about it. The late Neil Armstrong, the first man to walk on the Moon, summed it up nicely: "It would have been harder to fake it than to do it."[53]

Only 5 percent of Americans know how many astronauts walked on the Moon. Most think it is only one, two, or three.[54] The vast majority of people seem to think there was just one mission, Apollo 11, presumably. But the reality is that NASA made *nine voyages* to the Moon. Six of those landed, and a total of *twelve men* walked on the surface. This raises the question of why the government would have gone to all the trouble of faking it so many times. Every new filming session on a movie set somewhere would have increased the chances of being caught.

Perhaps if Moon-hoax believers were familiar with all of the Apollo missions and understood something about the critical preliminary Mercury and Gemini work of figuring out how to launch, orbit, spacewalk, rendezvous, and live in space, they would see this in a new light. No one just announced to the world out of the blue that men were on the Moon. Thousands of people put in years of very hard work to make it happen. This was one of humankind's finest moments, one *everyone* should share in.

NOSTRADAMUS

Excluding purely religious figures, Michel de Nostredame, or Nostradamus, is easily the most popular supernatural/paranormal predictor of future events in all of human history. This guy died in 1556, and yet today he continues to pop up in new books, websites, and TV shows. The History Channel seems committed to making sure the Nostradamus legacy lives on with its onslaught of questionable documentaries that link his writings to everything from the war in Afghanistan to Area 51. The funny thing is, however, Nostradamus doesn't deserve any of this attention. His predictions, if we can even call them that, collapse under the cruel and unforgiving weight of about five minutes of thoughtful analysis.

Nostradamus was a sixteenth-century French doctor and astrologer. That mix of trades was not as crazy then as it might seem today. People in the 1500s were deeply ignorant about medicine and astronomy, so it made perfect sense to have your doctor explain how your birthdate determined the fate of your career and love life while he bled you dry in order to cure your sore throat.

Understandably, one might think that there must be something special about Nostradamus. After all, there were many fortunetellers, astrologers, and soothsayers banging about in Europe back in his day, but it's his name that we all know. Even now, there are plenty of people who make magical predictions all the time. Some of them have many followers and earn lots of money. Their track records seem no better or worse than Nostradamus, yet he towers over them. Why?

Nostradamus is known today for only two reasons: good marketing and bad thinking. The reason his name survives and thrives is because few people take the time to think critically about what he is supposed to have written and whether or not any of it can reasonably be considered accurate predictions of important future events. The weak skepticism that is endemic in the public has allowed his myth to grow against all odds. His fame comes from the snowballing momentum of a concoction long known to capture minds and win hearts. It's often described as "smoke and mirrors" or, if you prefer, just plain "bull." The Nostradamus phenomenon is a good example of the *illusion-of-truth effect*. Hearing something over and over makes it believable to many people—no matter how silly it may be. No doubt conformity or the *bandwagon effect* comes into play here for many people as well. If you happen to have a circle of friends who constantly rave about the remarkable powers Nostradamus had, then your fallible human brain might be lured in.

A crucial piece of the Nostradamus puzzle is the manner in which he went about communicating his predictions. He wrote them as quatrains, four-verse poems. If you ever want to make it as a prophet, make sure to be as poetically ambiguous and confusing with your words as you can. The more flowery and hazy your language is, the more likely your predictions will be hailed as correct by sloppy thinkers one day. For example, Nostradamus did not write, "A viral plague will sweep across the world in 1918 and kill more than twenty million people." He did not write, "On July 20, 1969, three men will journey to the Moon and two, named Armstrong and Aldrin, will walk upon its surface." Predictions like that are far too clear and easy to give a pass or fail grade to.

We can't even be sure that Nostradamus wrote the things his fans say he wrote because none of his original works have survived. The quotes you may have read or heard about on TV are translations of questionable copies of secondhand copies, and so on. Who knows? Maybe he really did predict the future with total accuracy, but we'll never know because we can't assess his original work. What we do know is that the predictions that have been popularized are not impressive by any reasonable standard. Read these lines attributed to Nostradamus and judge for yourself:

> Beasts wild with hunger will cross the rivers:
> Most of the fighting shall be close by the Hister,
> It shall end in the great one being dragged in an iron cage,
> While the German shall be watching the infant Rhine.[55]

Well, that couldn't be any more clear. Obviously it is referring to Adolf Hitler, which is precisely what countless Nostradamus experts have declared. "Hister" is very close to "Hitler," right? No denying that. Here we have a reference to someone who would not even be born until centuries after Nostradamus's death, and yet Nostradamus knew of him. But hold on, "Hister" may sound a lot like "Hitler," but it sounds even more like *Hister*, which is what the Danube River in Germany was called during Nostradamus's day. Another clue that this might be about a river and have nothing to with a twentieth-century dictator is the fact that the line preceding it says something about crossing a *river*. Don't think I picked out a bad example to make Nostradamus belief seem silly. This is one of the most popular quatrains of all. No doubt some Nostradamus believers do not agree with the Hister-Hitler connection. Without exception, however, every believer I have had discussions with cited this one in attempt to convince me that Nostradamus had supernatural powers.

There may not be much to the Nostradamus myth, but there is a valuable lesson here that can be applied broadly. If we don't want to be duped, we have to resist placing trust in strained interpretations and trying too hard to make a claim fit the facts.

Good skeptics also give thought to context. Just because someone may be on a fancy television show that has nice special effects and a pleasing music score doesn't mean that she necessarily knows what she is talking about or is telling the truth. Too many people hear a sound bite from a UFO, ghost, or Nostradamus "expert" and feel inclined to accept it as fact. Don't do that! Think less about *who* is making a big claim and more about *what* is being claimed. Even if a Nobel Prize–winning scientist were to announce tomorrow that after careful study she had come to the conclusion that Nostradamus was a genuine prophet with supernatural abilities, it wouldn't be enough to convince me. She would get my attention based on her credentials, of course, but without very good evidence, her claim would mean nothing.

NEAR-DEATH EXPERIENCES

One of the most popular beliefs of all time is the one that says we get to live on in a different form and in a different place after we die. If true, it means that death isn't really death in the sense that it marks the true end of it all. It's a transition, nothing more. This is an extraordinary claim, of course. Unfortunately, it doesn't have any extraordinary evidence to go with it. The best we have are unusual stories from a small number of people who claim to have died and then been whisked away to heaven or hell for a brief visit before returning to Earth to live again.

Near-death experiences are viewed by many people as nothing less than absolute proof that there is an afterlife. I certainly agree that these accounts can be powerful and make a huge emotional impact on us. I met a woman who said she died and visited heaven before being sent back to Earth by Jesus himself. She told me all about the gates of heaven being "fixed with pearls and diamonds," streets that were literally paved with gold, and fruit bowls that magically replenished themselves when emptied. She saw angels teaching newcomers how to properly worship God. And then she met Jesus. "His beard was cut perfectly," she said. "And he had the most beautiful, crystal-blue eyes. He spoke to me, saying he would anoint me and send me back to Earth."[56]

I also have read books by people who claim to have died and gone to heaven. I even read one by a man who says he died and went to hell for a brief time. When people cite stories like this in an attempt to convince me that there is an afterlife, however, I point out that although there may be an afterlife, we can't rely on *stories* to prove it. This is too big and too important to simply trust a tale told by a person who is no less fallible than the rest of us. I may be willing to assume that a particular storyteller is sane and honest, but how can I be sure that this person did not simply experience an extreme dream that left him with an emotionally charged memory that he then confused for a real experience? We know that this kind of thing can and does happen. The man who went to hell, for example, began his amazing journey from his *bed* after going to *sleep* at night. He woke up on the floor next to his bed. He was shaken up and scared, and says he clearly remembers being in hell. But did he really ever leave his bedroom? Aren't a dream and the imperfections of human memory the most likely explanations for what happened to him? Without evidence, I would never accuse any of these people of being dishonest. But I do accuse all of them of being human. And with that comes an inescapable vulnerability to wild dreams, delusions, hallucinations, and flawed or false memories.

Brain science can provide likely natural explanations for the standard components of most near-death-experience stories. A light at the end of tunnel, feelings of peace and tranquility, seeing loved ones, and meeting religions figures can all be explained by natural processes in the stressed or dying brain. Nothing supernatural is necessary. This does not disprove the existence of some kind of an afterlife, of course, only that these near-death-experience stories are not the overwhelming proof so many people think they are. If only some of these people could bring back a piece of heaven to show us and prove their claim. Maybe a magically replenishing fruit bowl, for example. If physical objects can't make the return trip, what about information, something so special and unique that we all would know that somebody really did go to heaven or hell? But all we ever hear are dramatic stories that easily could have been produced by dreams and imagination.

When the brain is short on oxygen, very weird things can happen. Pilots report having tunnel vision (light at the end of a tunnel) before blackout from extreme g-forces in a cockpit or during a centrifuge ride. This is because blood drains from the eyes first before blackout. People who are not dying but who have brains that are being stressed and oxygen-deprived also report seeing people and things that are not there. Scientists can even induce hallucinations and out-of-body experiences in a person who is *not* oxygen-deprived or dying by simply targeting a specific area of the brain with a mild electrical current. That's it. No afterlife needed, just electricity.

I can personally connect with stories that describe feelings of pure peace and calm even as death approaches. When I was around twelve years old or so, I banged my face and cut my leg during a bad fall while crossing a canal on an elevated sewer pipe. Battered and bleeding, lying on the canal's bank, I was in bad shape. But I felt great! My brain took me someplace far away from agony and panic. I remember lying there, face in the dirt, feeling nothing but groovy. It wasn't bad enough to qualify as a near-death experience, and I don't remember feeling like I ever left my physical body, but it was definitely an unusual experience. The nice feelings soon faded, of course, and pain arrived. Because of that incident, I have no problem understanding what people are talking about when they say they felt happy and wonderful even though they were dying. I also can believe people when they say they had unusual visions or felt like they went somewhere in this state. But why would anything like this be proof of something so extraordinary as heaven, hell, or some other kind of afterlife?

Maybe it would help if we all keep in mind what the human brain is capable of conjuring up when things are going well. On a nice day, when the Sun is shining and our health is optimal, we are all capable of being fooled by our brains to see, hear, feel, experience, and remember things that are not real and never happened. If this is possible—and we know it is—then shouldn't we be skeptical about claims of experiences that occur to people who are near death or thought to be "dead" briefly? Don't you think a human brain under that kind of stress would be even less reliable?

If one wants to *hope* for something more after death, that's fine, of course. Like most people, I certainly would love to cheat death somehow. But no one should be misled to think that there is some vast collection of near-death cases that have proved the existence of an afterlife. All we have are stories, which alone prove nothing.

BERMUDA TRIANGLE

There is a mysterious zone over a patch of sea that functions a some kind of a natural or supernatural black hole, sucking up ships and planes that are never to be heard from again. Is it a portal to another universe? Could there be extraterrestrial involvement? Is the lost city of Atlantis involved? All we know for sure is that something very strange and very dangerous is happening in the area known as the Bermuda Triangle. Wow! What an exciting claim! I have lived much of my life very close to the Bermuda Triangle. I've even gone swimming in it. How exciting! Too bad it's nonsense.

One way to better understand and evaluate a weird claim such as the Bermuda Triangle is to look at its history. Where does it come from? Who was the first person to make the claim? What was the original context? How was the case for it first made? Finding the answers to such basic questions often sheds enough light on the belief to expose it as unworthy of your time. As we will see shortly, there are the usual problems with evidence and interpretation that slay the Triangle myth. But first, its origin makes it clear that this claim was hopelessly lost at sea from the start.

The earliest known mention of the Bermuda Triangle was an article in *Argosy* (February 1964). This was a pulp-fiction magazine that published fictional stories designed to excite and entertain young males. That's right, it wasn't the US Navy, the US Coast Guard, or *National Geographic* that broke the story and first alerted the world to the dangers of the Bermuda Triangle. No, it was a magazine that specialized in *adventure fiction*. The article's author, Vincent Gaddis, was probably looking for nothing more than another freelance check and never imagined he would help to create one of the great myths of modern times. But he did.

A writer named Charles Berlitz outdid Gaddis and really cashed in on the Triangle with his "nonfiction" bestseller *The Bermuda Triangle* in 1974.[57] Against all reason, it was such a hit that it anchored the Bermuda Triangle in pop culture by convincing millions around the world that the threat was real and deadly. Sadly, it didn't matter that he failed to present a convincing case based on facts and logic. Berlitz's "evidence" consists of nothing more than a collection of outrageous, unproven claims and heavily embellished sea tales. Berlitz also wrote other books, such as *Atlantis: The Eighth Continent* and *The Roswell Incident*, whose "evidence" is of similarly questionable accuracy.[58]

For a comprehensive takedown of this belief, read Larry Kusche's excellent book, *The Bermuda Triangle Mystery—Solved.*[59] Kusche dismantles claim after claim and shows how a minor story, or a made-up story, can be twisted and embellished to seem like compelling evidence of something sinister and supernatural. For example, the tragic incident of Flight 19 is probably the most famous of all Triangle stories. According to Berlitz and other Triangle proponents, however, the five US Navy Avenger torpedo planes took off from Fort Lauderdale airbase on December 19, 1945, and in ideal flying conditions vanished without a trace in a way that defies any natural explanation. Kusche shows that an honest look at the facts reveals a different story, however. Flight 19's commander was new to the base and unfamiliar with south Florida. He probably got his group lost, and they ran out of fuel before they could find land. They then ditched at sea, in the dark and in rough sea conditions. It should surprise no one that no wreckage was ever found. Heavy things like airplanes tend to sink in the ocean. Suddenly it doesn't seem quite so paranormal, right?

This leads us to the greatest of all problems with the Bermuda Triangle claim: When planes and ships travel over water, bad things sometimes happen. We should expect this. Due to the nature of accidents at sea away from land, there are not always clear and obvious explanations available. But a mystery or unanswered questions is not proof of a supernatural or paranormal event. This is the problem with the Bermuda Triangle. People have attempted to explain many different incidents with one overall theory. But there is not one cause, supernatural or natural, for thousands of mishaps over many decades in this one area or any other. "My research, which began as an attempt to find as much information as possible about the Bermuda Triangle, had an unexpected result," writes Kusche. "After examining all the evidence I have reached the following conclusion: *there is no theory that solves the mystery*. It is no more logical to try to find a common cause for all the disappearances in the Triangle than, for example, to try to find one cause for all automobile accidents in Arizona. By abandoning the search for an overall theory and investigating each incident independently, the mystery began to unravel."[60]

I'm confident that ships and planes do not vanish in the Bermuda Triangle at a rate or in a manner that indicates there is anything unusual going on there. But I like to be thorough, so I checked with both the US Navy and the US Coast Guard, just to be safe. Neither organization seems worried about this. The US Navy has more reasons than anyone to be concerned about threats to ships and planes. After all, it routinely sends thousands of sailors and aviators into the zone, along with billions of dollars' worth of ships, submarines, and aircraft. Yet the navy has gone on record rejecting any consideration of the Bermuda Triangle as an unusual threat to them.[61]

Finally, if anyone should know if there is anything to the Bermuda Triangle claim, possibly even more so than the US Navy, it would be the people who are specifically responsible for safety and rescue in and around the Triangle area. So I phoned the Miami station of the United States Coast Guard and spoke to an officer there who told me that they do not believe in the Bermuda Triangle story, nor do they give it any consideration when going about their duties. He also referred to this official statement by the Coast Guard:

> The Coast Guard does not recognize the existence of the so-called Bermuda Triangle as a geographic area of specific hazard to ships or planes. In a review of many aircraft and vessel losses in the area over the years, there has been nothing discovered that would indicate that casualties were the result of anything other than physical causes. No extraordinary factors have ever been identified.[62]

Case closed. Mystery solved. Let's have a long-overdue burial at sea for this one. The Bermuda Triangle needs to go back to being what it originally was: just a good story meant to entertain and not to be taken seriously by serious people. It's a great campfire tale to tell the kids while vacationing in South Florida or the Caribbean but nothing more.

Sadly, however, after appearing to fade in the 1990s, the Bermuda Triangle is enjoying a bit of a renaissance today. No doubt this is due to the abundance of misleading websites that promote its existence as historical truth and scientific fact. There has also been an avalanche of terribly misleading pseudo-documentaries about it on cable TV. These have modernized and popularized the myth for a whole new generation of uncritical thinkers.

ATLANTIS

Atlantis, the extraordinary ancient city/continent that was struck by disaster and sank into the sea is an example of an unproven belief that may be more fun than threatening. So what if some people believe in a mythical place that was more technologically advanced than we are today and spawned all the great ancient civilizations? What's the harm?

The problem is that Atlantis belief is a symptom of something more serious. Falling for this claim can be viewed as a wailing alarm that indicates something is wrong with a person's thinking process. If Atlantis belief makes it into one's head, then it means that one's skepticism force field is either switched off or turned down way too low. After all, if one is capable of believing this claim—despite all of its logical problems and a total absence of proof—then it's likely not going to take much of a leap for this person to believe in a potentially dangerous medical product, invest in scams, or be exploited in a variety of ways by unscrupulous people and organizations. Alone, such beliefs may initially appear to be harmless. But seemingly benign claims may lead people deeper into a variety of weird beliefs and delusions that do cause harm. For example, Atlantis belief has an odd connection to unproven New Age claims about crystals having healing powers. How easy would it be for some people, encouraged by their Atlantis belief, to begin wasting time and money on crystals? This could easily lead to serious problems if people choose crystals over medical science. A "harmless" belief in Atlantis could become a very dangerous belief in the medical value of overpriced rocks. By the way, some readers may suspect that Atlantis is a fringe belief in the extreme, one not worth addressing because it involves so few people. According to surveys, however, this is not so. A 2006 Baylor study, for example, found that more than 40 percent of American adults believe in "ancient advanced civilizations such as Atlantis."[63] More alarming than that, however, is another study that found 16 percent of high-school *science teachers* believe in Atlantis.[64]

As is so often the case when evidence is not involved, the description of a belief varies wildly and depends on which believer one asks. Some Atlantis believers say it was a prominent city/continent that existed several thousand years ago and was advanced far beyond its ancient peers and, some say, even beyond twenty-first-century civilization. For example, some claim that the Atlanteans had nuclear weapons, aircraft, and energy-production technology that was far superior to ours today. Some believers go so far as to claim that extraterrestrials were/are involved with Atlantis either as visitors or were the Atlanteans themselves. Virtually all believers agree that the city/continent vanished due to a catastrophic event such as an earthquake or a tsunami. Today it is supposed to rest on the floor of the Mediterranean Sea, at the bottom of the Atlantic Ocean, or maybe under antarctic ice somewhere, awaiting discovery. Another popular component to the myth is that before its violent end, Atlantis spawned satellite cultures that were the seeds of modern civilization. One version says we are the descendants of inferior slaves who escaped just before disaster struck the city. Either way, all people today are deeply connected to Atlantis. Okay, I admit it, this is another great story. I get why people might feel an urge to believe it. But if you think first, it's not quite so impressive.

Before we consider some of the reasons why doubt is in order, let's be clear about something important: *Atlantis has not been discovered, and no compelling archaeological or historical evidence for it has ever been produced by anyone.* This needs to be stated because many people seem to think it has been found or, at the very least, there is enough evidence to confirm that it did exist. Many people have likely been misled about the status of Atlantis because its discovery or near-discovery is reported frequently by the mainstream news media. Consider these recent headlines: TSUNAMI CLUE TO ATLANTIS FOUND[65]; SATELLITE

IMAGES "SHOW ATLANTIS"[66]; ATLANTIS "OBVIOUSLY NEAR GIBRALTAR."[67] I chose these headlines specifically because they all come from the BBC, one of the world's most respected and trustworthy news organizations. In fairness, the BBC did place quote marks around key words within two of these headlines. This is to shift the burden of the claim off of the BBC and onto some person who made the claim. Also, in many cases, reports like this usually do reveal somewhere in the text of the article that nothing of consequence has been found and absolutely nothing about Atlantis has been confirmed. These kinds of reports have always turned out to be nothing more than an "exciting lead," a "possible clue," or "encouraging data"—anything but actual artifacts that confirm the existence of Atlantis. The damage is done, however, because many people see headlines like these and don't bother to read the entire article, or, if they do, can still be deeply influenced by the eye-catching headline. If they don't quite turn into confident Atlantis believers, they still might come away with the false impression that there must be *something* to this claim. But, as we are about to see, there is nothing here but a tall tale.

The only references to Atlantis in ancient times come from an impressive source: Plato, one of history's great philosophers. He mentions the city in two of his dialogues, the *Timaeus* and the *Critias*. But what did he actually communicate about Atlantis? It seems that for the purposes of teaching a lesson, Plato referred to an old "story" about a city-state that was too militaristic and too arrogant for its own good. This suggests that he wasn't presenting a serious historical report on a real place and real people. He was almost certainly using a fictional tale to help make a point to the people he was addressing. There is nothing about Plato's mention of Atlantis that makes the case for its existence. This view is supported by the fact that there is total silence about Atlantis from the rest of the ancient world. Atlantis was supposed to have been vastly superior in technology, science, and culture, and to have given rise to all other great civilizations, yet there is no mention of it from the Egyptians, Carthaginians, Romans, Sumerians, Babylonians, Ionians, Macedonians, and so on. Archaeologist Kenneth L. Feder hammers this point home in his excellent book, *Frauds, Myths, and Mysteries: Science and Pseudoscience in Archaeology*: "You will not read the discourses of *modern historians* arguing for or disputing the historicity of *The Lord of the Rings* or *Harry Potter*, because these are understood, of course, to be works of fiction. In much the same way, Greek historians who followed Plato did not feel the need even to discuss his story of Atlantis; they understood it as the work of fiction Plato intended it to be."[68]

An important thing to keep in mind is that believers often try to lure skeptics into arguing against things they don't even oppose. For example, I have had discussions with Atlantis believers who, after becoming frustrated by their failure to convince me, changed course and accused me of being close-minded, antiscience, and unwilling to consider the possibility that underwater archaeologists might find an ancient city one day, maybe even one that was called Atlantis. Of course they *might*. I am a big fan of underwater archaeology. I have interviewed Bob Ballard, discoverer of the *Titanic* and the Galapagos Rift hydrothermal vents, and I share his enthusiasm and optimism for underwater archaeology's potential. We have been a coastal-dwelling and seafaring people for a long, long time. Lakes, seas, and oceans undoubtedly hide many amazing artifacts that can shed new light on our past. Furthermore, given the volatile nature of our world and its changing shorelines, the claim that an ancient city is lying in ruins underwater somewhere is not only reasonable, it's very likely to be true! Who knows? Maybe there is even a city that was named "Atlantis" out there. But until somebody finds it, we don't know. And if the ruins of such a city are found, that still would not prove that it had once been inhabited by an advanced race of people or some alien species. To stay true to the real world, we have to ask the right questions, hold out for evidence, and not allow ourselves to be sidetracked by claims we're not refuting—in other words, we need to try to become and remain good skeptics.

AREA 51

Imagine you are the president of the United States or a top Air Force general. A frantic aid has just burst through your office door. "My God, it's been confirmed! Our team on the ground has secured the crash site and report that it is definitely an extraterrestrial vehicle of some kind, and the deceased bodies of four alien occupants have been recovered. What are your orders?"

You would have the wreckage and the bodies sent to Area 51, of course. Where else? Everyone knows that's where the government stashes all visitors from outer space. Area 51 shows up in so many science-fiction books and films and has been a flashpoint for UFO believers for so many years now that is has evolved into something of a comical cliché in pop culture. I won't be surprised if I turn on my TV one day soon and hear something like this: *Lee Harvey Oswald, an alien-hybrid Libra with psychic powers, teleported to a hangar inside Area 51 where his brain was reverse engineered by former Nazi scientists who themselves had been chemically lobotomized by the Men in Black who were actually rogue CIA agents in close communication with fluoride-free extraterrestrials.* How did Area 51 become ground zero for weird American myth? What is Area 51? And what really goes on there?

First of all, Area 51 does exist and mysterious things do happen there. No one can deny that. After many years of secrecy and denial, it is now a matter of public record that very unusual things have been going on at this top-secret government base for decades. But aliens on ice? Flying saucers? Reverse-engineered extraterrestrial technology? Probably not. Believers in Area 51 mythology get tripped up when they confuse plain *secret aviation activity* for *secret aviation activity involving aliens*. They also mistake *secret military aircraft* with *extraterrestrial spaceships*.

For such a big secret, an awful lot is known about Area 51. We know, for example, that it is part of a larger US military base located in the Nevada desert, less than ninety miles from Las Vegas. The general area is also called Paradise Ranch and Groom Lake. The US military develops and flies new aircraft there. Many of these flights are conducted at night to enhance secrecy. The CIA also uses the base for the same purpose. One would have thought that "UFO sightings" around a region like this would be expected by everyone, not met with suspicion and runaway imagination.

The list of aircraft now publicly known to have flown at Area 51 includes some revolutionary planes. Amazing flying machines like the U-2, A-12, and the now-famous SR-71 Blackbird flew there in extreme secrecy from the 1950s through the 1990s. These sleek reconnaissance planes were extraordinary both in appearance and for what they could do. They were capable of flying at speeds faster than 2,000 mph and reaching altitudes above 85,000 feet. By comparison, the cruising altitude for a typical modern commercial jet is 35,000 feet, and they usually fly at speeds of less than 600 mph. The SR-71's Cold War mission was to fly far, high, and fast in order to take photos of Soviet bases and activities on the ground. Even today, it's difficult not to be impressed by the sleek and futuristic lines of the SR-71. There is one on an elevated display at the San Diego Air and Space Museum that allows visitors to walk under it. The underbelly view makes it very easy to imagine how a layperson fifty years ago might have felt certain that she had seen something from another world if one of these roared overhead shortly after takeoff or was spotted out of a window high above a "normal" plane in flight.

The B-2 Spirit bomber and the F-117 Nighthawk also flew at Area 51 before anyone outside of select military and government channels knew anything about them. The B-2, with its sweeping batwing and tailless design, looks like something a Hollywood special-effects company might have come up with. Seen in flight, from some perspectives, it does not appear to be a conventional plane at all. It can look—surprise!— very much like the stereotypical flying saucer. The F-117 Nighthawk is another bizarre warplane. It has a boxy design with unusual angles and an unexpected rudder-elevator-fusion configuration in the rear. We've come a long way from the Wright brothers. How far is anyone's guess, however, because the strange aircraft that we *know* flew at Area 51 can only tease us and hint at what we don't know and what might be flying in secrecy there or at some other base right now.

Some people cite what they say is a high number of UFO-sighting claims around Area 51 as proof or evidence that the government has alien spacecraft there and is operating them or has reversed engineered alien wreckage and built its own. The first problem with this is that seeing unknown aircraft near a secret base that flies secret aircraft is not strange. Another problem is one common to many extraordinary claims like this: unanswered questions are not answers. Not being able to identify or explain what something in the sky is does not warrant jumping to the amazing and outrageous conclusion that it must be a spaceship of extraterrestrial origin. Some people are very critical of the US military for its secrecy and refusal to give straight answers about what it is up to. But what else should we expect? Secrecy is to be expected from *secret* military and CIA operations. It certainly is not proof of anything having to do with aliens.

Keep in mind that many years may pass between the time new, secret aircraft are first tested at a base like Area 51 and when the existence of that aircraft becomes known to the public. This is how it has been in the past, so why wouldn't it be true today? I have no doubt that there are strange things being flown in the night skies over America right now, vehicles that would shock those of us who are out of the loop. I suspect that this may be truer today than ever before because of the explosion of UAVs (unmanned aerial vehicles, or drones) and the impact of the post-9/11 "war on terror." Tremendous amounts of money and creative thought are being poured into making new, better, and increasingly exotic aircraft. Smithsonian's *Air and Space* magazine reports that an estimated 30 to 36 billion dollars per year is being spent on new aviation technologies specifically at Area 51 alone.[69] That works out to be as much as $100 million *per day*. The UAVs or drones we know about today are weird enough, so imagine all the bizarre designs that are likely being tested right now. At the very least, you can be sure that some very interesting work is going on at Area 51 or somewhere else that we don't know about. And if you ever catch a glimpse of some of that work in flight on a dark night, you just might suspect that it is something from another world. But that hunch would not make it so.

EXTRAORDINARY RELIGIOUS CLAIMS

It would be easy to sidestep religion and just stick to raising questions about less contentious topics such as astrology and UFOs. But good skeptics don't give a free pass to anyone or anything. Good skepticism demands consistency. It does no good to figure out that giving thousands of your hard-earned dollars to a psychic is a bad idea only to go running into the open arms of a faith healer who wants you to help pay for his fleet of luxury cars. I understand that religion tends to be far more serious to most people than many of these other claims, but doesn't this greater importance indicate a need for *more* vigilance rather than less? If gods, sacred books, and heaven and hell really are crucial to our well-being, then why would anyone want to risk worshipping the wrong god, following the wrong rules, reading the wrong book, or trusting the wrong prophet? It's complicated, to say the least, and skeptical thinking can help to sort it out. Over the last several thousand years, people—most of them smart, well-meaning, and sincere—have confidently believed in millions of gods and created hundreds of thousands of unique religions, most of them too contradictory to ever be reconciled logically. It is not rude to encourage everyone to think critically about this situation. If anything, it's demanded by what we see in human belief today and in our past.

Many people who hold to one form of religious belief or another are pretty good skeptics already. They see right through the cons, distortions of history, logical errors, and outright lies that infect so many religions, including their own. There are numerous believers, for example, who are unflinchingly skeptical of the work of some millionaire holy men and consider them to be nothing more than liars and thieves who prey on believers who are not skeptical enough. Millions of believers think well enough to reject the teachings of leaders who insist that every storm is a divine punishment and every lucky break is a gift from the gods. The point is, skepticism should not halt at religion's front door. It is appropriate, useful, and vital *even for those who may feel they will not or cannot give up their core belief in a god or gods.*

I suspect that far too many religious people shy away from skeptical thinking, or even feel hostile toward it, because they see it as a direct attack on the most important belief of all. But when religious people turn away from critical thinking, what happens? They leave themselves open to problems they don't want. For example, how many millions of religious people have fallen victim to medical quackery and other scams or wasted their time and money on bad beliefs that they could have identified and rejected if they had been better skeptics? The answer is *too many.* I don't hold back from encouraging religious people to be good skeptics because they need it, too. They have to live in this crazy world, too. Religious people need skepticism as much as anyone. I certainly don't care about them any less than I do nonreligious people. It's so important, in fact, that I encourage religious people to become good skeptics even if they insist on being selective about it. That's not ideal, of course, but it's better than not trying at all. Even if one decides that it is necessary to leave one's particular god or gods untouched by critical thinking, at least apply vigorous and consistent skepticism to every other aspect of life.

Skepticism is important for religious people *within* their religion because wrong ideas and bad beliefs can sprout up there, too. The question of the existence of gods aside, religions often provide a protective bubble that nurtures and protects ideas nobody needs banging around in their head. This is not meant to insult any particular religion; it's just reality. Everyone knows that religions, like any organization, can be used to motivate, excuse, and protect both self-destructive and socially destructive behavior. Consider the suffering of hundreds of millions of women who are mistreated every day, worldwide, by men who believe that such behavior is what their religion expects of them or that such treatment is acceptable because their religion allows them to get away with it unpunished. Think about the countless children who are taught outrageously bad pseudohistory and pseudoscience by parents and teachers who think it's their duty because of a religion they adhere to. Creationism, for example, includes the profoundly unscientific claim that our world is only about 6,000 to 9,000 years old. But the Earth is around 4.5 *billion* years old. So this claim is not just wrong; it's *stunningly* wrong. It's no less silly than saying that the distance between the Earth and the Moon (approximately 240,000 miles) is about one-third of a mile. Or that the distance between New York and Los Angeles is about twenty feet. Honestly, that's how far off the mark this popular claim is. Although you would expect that no educated person in his or her right mind would believe in a 6,000-year-old Earth in the twenty-first century, in fact, many millions of otherwise-sensible people do because it's a bad idea that thrives under the protective dome of some versions of some religions. Religious people who also happen to be pretty good skeptics, however, would easily identify problems with this claim and reject it. They would ask a few simple questions, note the convergence of so much evidence from astronomy, physics, paleontology, paleoanthropology, archaeology, botany, biology, and geology, and conclude that the Earth is obviously much, much older than 6,000 years. The end result would be a more enlightened person who still believes in his god or gods but is not clinging unnecessarily to a pseudoscientific claim that makes him less aware and less appreciative of the real world and universe he lives in.

Skepticism is good for everyone at all times. There should be no religious zones that are off-limits to asking questions and requesting evidence. Religious people are no different than everyone else in that they are confronted with lies and popular delusions every day. Sometimes this happens within the bounds of their particular religion. Hopefully believers will agree that all religions are imperfect because they are organized and led by imperfect human beings. Every religion has taken wrong turns somewhere in its past, and every religion has had some leaders who turned out to be underserving of their followers' trust. For these reasons, it only makes sense for religious people to embrace and apply skepticism for their own protection.

GOOD THINKING!

- Good skeptics do not claim to know with absolute certainty that all unusual and unproven claims are not true and could never possibly be true. A very important part of being a good skeptic is maintaining an open mind.

- A good skeptic doesn't stubbornly defend a rigid position against a weird claim. The goal is to base sensible conclusions on the best available evidence and arguments. If truth and reality were to translate to a world with ghosts, magic crystals, and vampires in it, then that is the world good skeptics want to know about.

- Skeptics are often accused of being "against everything" when in reality it is only mistakes, delusions, and lies that they are opposed to.

- It helps if skeptics can do more than just demand evidence when faced with an unusual belief. Try offering ideas that are relevant to the claim, a bit of historical context, some science, some insight into how the brain works, and maybe a good alternate explanation. Good skeptics encourage people to think critically about ideas; they dismantle bad beliefs, not believers.

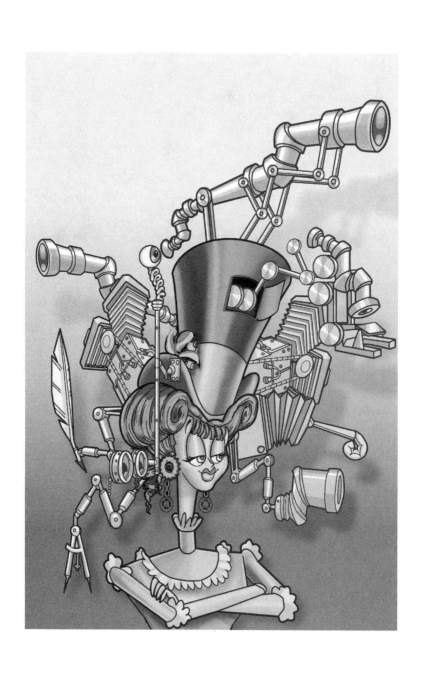

Chapter 4

THE PROPER CARE AND FEEDING
OF A THINKING MACHINE

D o you love your brain? I mean really, really love it. Do you treat it as well as you can? Do you ever pause to think about how much you need this three-pound, convoluted contraption that is squeezed into your skull? Do you appreciate and admire the way it keeps you going, day after day? Do you feel grateful for how it got you through that tough math class in middle school? Did you give it any credit at all for serving up just the right words at just the right time when you finally made a move on your high-school heartthrob? Did you remember to say thanks when you successfully maneuvered your way through and around workplace politics to earn a raise or promotion? And when you wake up every morning and discover that you didn't die during the night, are you thoughtful enough to thank your brain for keeping an eye on things for you while you were checked out? After all, it was your brain that made sure to keep you breathing all night.

If you are like most people and don't think much about what your brain needs to be healthy and do its job well, you need to ask yourself why. What sense does it make to give little or no thought to the center of your own personal universe? Your brain is the beginning and end of you. There is no other organ like it. Not even close. Your brain is the headquarters, the theater, the museum, the love zone, Grand Central Station, ground zero, the control tower, the national archive, the library, the music studio, the well of all emotions, the source of all drive and determination, the place where courage and fear duel for supremacy. It's *everything*. The rest of you is pretty much just plumbing and hardware. The brain is where all the magic happens. There is a reason we don't hear the phrase "heart dead" but do hear "brain dead." Your brain's life is your life. Other body parts, you can fix, replace, or do without. Missing limb? Prosthetic, please. Bad heart? Transplant time. A dead brain, however, means you are over. You can't replace it because it is *you*. Perhaps there will be a day in the future when we can upload the information in our brains to hard drives for safekeeping or to function independent of our bodies, but for now this is the only thing you have to hold your personality, instincts, memories, dreams, and learned abilities. In light of all this, I ask again: Do you treat your brain as well you should?

If not, why not?

BRAINS DESERVE BETTER THAN THIS

The truth is, most of us mistreat and neglect our brains. We take them for granted and act as though they are little more than cranial stuffing along for the ride. Maybe we would do better if we all had transparent heads allowing us to see the brains of other people and our own when we look in a mirror. Imagine if we could recognize healthy brains and battered brains when they passed us on the sidewalk. Perhaps then brains might get as much attention as, say, waistlines, hair, and fingernails do. Unfortunately, most of us tend to acknowledge the brain's existence only when it screams out in discomfort, and even then, many of us react in less-than-optimal ways. When it hurts because it's thirsty or hungry—reach for a pill. When it's sleepy because it needs downtime—soak it in caffeine. When it flounders and fails because it lacks the necessary information or training to do something we want it to do, don't admit the obvious and recognize that it needs more information and more training. Just assume that it's inherently weak and too genetically limited to do any better. *It's not my fault; it's my damn brain's fault. I botched that conference call to the sales team because my brain was foggy, not because I didn't sleep much last night. I didn't know the answers on that French Revolution midterm because I'm just too dumb—it wasn't because I didn't study.*

EMBRACING THE BRAIN

Our brains are too important, too special, and too wonderful to excuse such neglect and mistreatment. Therefore, I encourage you to force-feed yourself an epiphany right now. Embrace the brain. Fall in love with your thinking machine. Don't wait for dramatic background music or the rumble of distant thunder to mark the moment. Sorry to break it to you, but your life is not a movie. Just decide to do it. Commit to doing a better job of appreciating and caring for your brain from this moment on. It's the least you can do, considering all it does for you, every minute of every day. If your brain had taken just one ten-minute break at any point in your life, you would be dead right now. We all care about the beating of our hearts, but it's the buzzing of the brain that matters most of all. Try to remember that your brain doesn't only assist you in watching forty hours of television per week and finding the best junk food in large and confusing supermarkets. It also does a bunch of other things, too, chores like regulating your breathing, heart rate, blood circulation, digestion, and so on. It's always on the job for you, whether you thank it or not.

If you need help appreciating your brain, consider what came before. Think about what was necessary yesterday for you to exist today. You and your brain stand at the end of a long line of brains that lived and thought across many thousands of years. Hundreds of thousands of years ago, *Homo erectus* crafted stone tools and tamed fire. Everything accomplished in our civilization since shares a direct connection to the early achievements and resilience of human brains. Without them, there is no civilization. Go back even further, millions of years. Now imagine the small families of Australopithecines that survived the immense challenges of prehistoric Africa, not with superior teeth, claws, speed, strength, smell, or eyesight, but with superior *brains*. They didn't punch, kick, and bite their way to success. They reasoned and imagined ways of solving problems so that they could stay alive long enough to push their genes forward, genes you now carry.

According to the best current evidence, the first anatomically modern humans roamed the Earth some 200,000 years ago. Their brains, essentially the same as ours, gave them all the creativity and processing power they needed to endure, spread, innovate, and build. Try to sense and appreciate the link between you and all the people who lived long ago. You have a brain like that, a brain that conquered the world! They invented spears and then teamed up to hunt twelve-foot-tall cave bears and eight-ton mammoths. They sat around fires at night and told great stories. They used their brains to invent new ways to elude predators, catch prey, and improve their lives. They looked at the world around them and then reinterpreted it with another invention called art. They wondered what might be beyond the horizon and then set sail to find out. They imagined, designed, and then built great stone structures such as Göbekli Tepe, the pyramids at Giza, and the Parthenon. Their brains even came up with a new way of thinking, called science, that allowed them to discover and understand more of the natural universe. Thanks to the brains that came before you, you belong to a very exclusive club today. That big, bad brain you have been walking around with all of your life is the reason you are not cowering up in some tree right now, hoping the pack of wolves below gives up and moves on. It's the reason you are able to look up at the night sky and have some realistic idea of what you see and where you are. Because of your brain, you can set a goal for next year and imagine yourself achieving it. It's the reason you are something very special on this planet.

Not so touched by the challenges and triumphs of early humans? Just not feeling the love for that long, unbroken chain of brains that came before you? Fine, I've got an even better reason: Take up the proper care and maintenance of your brain because it is in your own self-interest to do so. Treating your brain with care can improve your work or school performance, athletic ability, romantic life, creativity, mood, sleep, physical health, and much more. If any of that interests you, here is the least you need to know . . .

EAT WELL, THINK WELL

One of the best things you can do to help your brain stay healthy and perform well is to *eat well*. Don't worry, I'm not going to get preachy here. This is not a diet book, and I'm not going to tell you to put down the ice cream. I only want you to know that good brains start with good nutrition. Bad eating habits can be devastating to the human brain. I have researched and written about global poverty issues for many years, and one of the most disturbing aspects of this topic is what poor nutrition does to intelligence and overall brain development. Unforgivably, our world squanders an immeasurable amount of intellectual potential every moment of every day because poor people, children in particular, don't get enough good food. But this is not exclusively a problem with those living in poverty. The connection between food and brains doesn't go away when income or Gross Domestic Product figures rise above a certain level. The biology remains the same. A brain is helped or hindered by the quantity and quality of its fuel, and this is true for every human. Chronic hunger is terrible for the brain, of course, but an 8,000–calorie per day, junk-food-based diet causes serious problems, too. Consider this. People are forever saying things like: *Thanks, Ma, but I think I'm done. I probably shouldn't eat another bowl of your lard pudding because it's bad for my heart.* Or, *I'm gonna ease off a bit on the fried butter this month. Don't wanna end up with a bum body that can't do nothin'.*

But we never seem to hear: *I'm skipping fast-food fries from now on because they're bad for my brain.* Or, *No more soft drinks for me, dude, I don't wanna end up with a brain that can't think well.* But this is precisely how we should be thinking. A nutritionally trashed brain is no less important than a body ravaged and bloated by poor food choices.

You need to know and never forget that your brain is very needy, perhaps even greedy. This organ that lives in your head demands some *20 percent* of the blood in your body to operate properly. This is a lot for an organ that weighs only about three pounds, which is less than 2 percent of a typical adult's body weight. If you are dehydrated, calorie deficient, or protein deficient, your brain will struggle. It will not think, create, solve problems, or be as alert as it can be. It will be more prone to bad moods. All this matters to the good skeptic, to anyone who wants to think well and think often, because if you mistreat your brain with bad eating habits, forget figuring out if it was the Loch Ness monster you just saw or a floating log; you won't be able to walk or do simple math as well as you otherwise could.

Try not to forget the obvious: Your brain is an intimately connected part of the overall system that is your body. It is linked to everything that happens to you. If you make the decision to consume too much of the wrong stuff (junk food, alcohol, high-sugar soft drinks), then your brain is going to feel that choice. Don't just worry about that second piece of cheesecake having eventual repercussions on your thighs and butt. Know that poor food choices show up in your brain, too.

It is unfortunate that nutrition in many societies these days is so often associated with, first and foremost, how fat one is. The first concern for most people seems to be how food affects the way we look rather than the way we think. Things more important than physical appearance like diabetes and heart disease may come to mind at some point, but brain health and performance is rarely thought of at all. Relatively few people maintain good eating habits or make the decision to improve their eating habits primarily because they want the best for their brain. But given what our brains do for us, they should be the first motivation.

So what should we eat to give our brains what they need to work well? There are plenty of unproven pills and potions that promise to make you a genius and hold off age-related diseases forever (you should know better than to fall for them by now). There are many compelling scientific studies about specific foods that seem to support both short-term and long-term brain health. I believe, however, that it is more productive for most people to start by adopting an overall philosophy of eating well for general body fitness and brain health rather than chase behind every new study that hints at the discovery of a new wonder food. Keep this simple, and there is a better chance it might work for you over the long run.

Try to do the basic and obvious good things, such as eating lots of vegetables and keeping the total amount of food consumed reasonable on a daily basis. The science concerning brains and good nutrition is now loud and clear. There is just too much for reasonable people not to react when they hear about it. For example, a study published in the journal *Neurology* found a strong link between the regular consumption of leafy, green vegetables (such as spinach) and slower rates of cognitive decline in old age.[1] But you already knew spinach was good and that you should eat more of it more often. You don't have to be perfect or view eating as a burden. The trick is to make smart choices over the long haul. Your habit and pattern *most of the time* needs to be smart eating. If you are a meat eater, then eat more wild salmon and less red meat, for example. Eat more baked chicken and fewer hamburgers. It's probably best to pass on hot dogs and sausage every time. Sugary soft drinks are never a good idea. If you must indulge, keep them to a minimum. For snacking, try to eat fruit and nuts more than candy. Find or create a trail mix concoction with quality ingredients that appeals to you and keep a bag of it handy throughout the day. I do recommend blueberries and blackberries specifically, as they appear to be good for the brain.[2] So stock up and start nibbling. Don't mistake healthy eating for magic. Consuming spinach, broccoli, and berries alone won't be enough to get you accepted into Harvard or make you cancer-proof. This is about making wise choices most of the time over a lifetime so that you at least give yourself a fighting chance to be fit, healthy, and smart.

Don't overcomplicate this or make yourself suffer. Just figure out which of the good foods you like best, then stock up on those. It's not a big deal. You should never be hungry, and you should never hate your meals. If that's happening to you, then you are doing it wrong. Take a moment to figure this out, make a few minor lifestyle and shopping adjustments, and you just might improve your eating (and thinking) habits forever. The goal is simple but profoundly important to the quality of your life: Eat well to build and maintain a body that is strong and healthy so that your brain gets what it needs for optimal daily operation.

STAND UP FOR THINKING

Eating well is extremely important for the brain, but it's not enough by itself. If you want the thinking machine perched on your shoulders to come anywhere near its potential, then you need to get moving. Based on very good research, we now know that exercise can have a significant positive impact on stress, anxiety, depression, and age-related problems such as dementia. We also know that exercise boosts brain performance for all ages, from the very old to the very young.[3] Exercise even causes your brain to *grow new cells*. This is amazing when you think about it: *Exercise grows your brain*. According to John Ratey, an associate clinical professor of psychiatry at Harvard Medical School, the brain benefits from physical activity primarily because of the increased blood flow that comes with doing things like walking, running, swimming, and cycling.[4] In the same way that exercise stimulates blood flow and delivers nutrients to muscles, it does so for the brain. Just think of your brain as a three-pound vampire; it needs blood to live—the more, the better.

Physical activity is a must. Again, I don't want to preach any readers into a corner and make them feel uncomfortable, but there is a hard and necessary truth to be told here: If you are spending virtually all of your waking hours sitting on couches, car seats, and office chairs, then your brain is not functioning at optimal levels. You are crippling it and wasting its potential with your physical inactivity. It's as if you are living your day-to-day existence with a pillow case over your head. You are here and you are awake, but you aren't really in a position to do your best at anything. The solution is clear: *Stand rather than sit when possible and move rather than be still when possible*. It's as simple as that. You need to be active and you need to sweat. Inactive and sweatless is a state better left to the dead. It's also important to put some stress on your bones and muscles in order to make and keep them strong. This is nothing personal. Nobody is being mean to you. I am not the ghost of your sixth-grade PE coach haunting you. This is a basic truth based on who you are. If you don't like it, then you picked the wrong species to be born into.

Like it or not, you are a member of a particular life-form that spent the last 99.999 percent of its existence on its feet and active. We walked and we ran, probably *several miles* per day. You are stuck with a body and brain that do not respond well to inactivity. *Homo sapiens* evolved on the go. We are not lichens or sloths. Virtually every one of your ancestors over the last few million years at least spent their days walking rather than sitting. This means the human brain evolved inside of mobile bodies. Because of this, our biology today works best when we are standing rather than sitting, and walking rather than standing. Every time you sit for hours and hours, you are rebelling against who you really are. And, unfortunately, it's a fight you can't win.

As with nutrition, there is no reason to feel the need to become obsessive about physical activity. If you want to run marathons and lift small cars, go for it. But don't be discouraged if things like glacier climbing and triathlons don't appeal to you. Thanks to science, there is very encouraging news for those who would just like to have healthy and highly functioning bodies and brains. It turns out that it's not so difficult to turn yourself into something that will pass for a fit modern human being. Combined with good eating habits, you can do it on as little as twenty minutes of moderate activity six or seven times per week. That's all! It can be walking, running, swimming, cycling, or maybe a mix of those and others. Just find what works for you and get busy. Try to include at least a couple of weekly weight-training sessions, too. And don't forget to eat well.

Many people who don't fully appreciate the connection between physical activity and brain/body health probably assume that they don't have the time for exercise and hide behind that excuse. *My life is just way too busy and hectic for me to find twenty minutes a day to improve my life. I'm so tired after thinking all day at school and work that I don't have any energy left to exercise in order to become more energetic and a better thinker.* The reality is that there is almost always time for exercise if one looks for it. For example, the average American spends nearly *forty hours per week* watching television.[5] That's like a full-time job by itself. Compared to that, walking or running and lifting weights for a total of just 140 minutes or so per week, barely more than two hours, is not a very big sacrifice of time and effort, especially when you consider the huge payoffs. There is no place for excuses here. Again, don't fail your brain because you equate exercise with training like a professional athlete and feel intimidated. Just find a way and start moving. It doesn't matter if you have sore knees, bad legs, or no legs. Virtually everyone can do something. If nothing else, do jumping jacks and push-ups in your living room during one thirty-minute episode of your favorite TV show every evening. Until you do, your brain will remain trapped in a fogbank. Set it free.

Age is irrelevant. Young, old, or somewhere in the middle, we all need to be active, or our brains won't be able to do their best work. According to a Centers for Disease Control (CDC) report about young students in school, there is now "substantial evidence that physical activity can help improve academic achievement, including grades and standardized test scores."[6] Researchers have also found that the benefits of physical activity for young students include "enhanced concentration and attention as well as improved classroom behavior." Regardless of income or culture, any brain that is lost inside an out-of-shape, sedentary, and poorly fueled body is thinking uphill and against the wind. There is no avoiding this; if you want to be a good skeptic with a sharp mind, then don't just sit there, get moving.

Just like it is with eating well, being physically active shouldn't be thought of as a burden. Try to make it fun, or at least as tolerable as you can. Then, before you know it, physical activity will turn into a habit. Exercising has been a part of my life for so many years, for example, that I'm hopelessly hooked. It's just part of who I am and what I do. I begin to feel terrible if I go more than three or four days without running or lifting weights. I also draw on some of the amazing research that has come from brain science for additional motivation and reassurance. For example, when I was in my twenties, I would skip working out if I couldn't find the time or couldn't summon up the energy for an epic workout that left me feeling like I had summited Everest. Now, however, thanks to science, I know that even just twenty minutes of walking or slow running helps my brain and body in numerous ways, so it's well worth it. Doing something is better than doing nothing. I have even taken to standing as much as possible when I write and read because it turns out that sitting for hours is bad for long-term health.[7] Brain scientist John Medina, author of *Brain Rules*, was so impressed by the research on the relationship between standing/moving and brain fitness that he put a treadmill in his office and attached his laptop to it so that he could work and walk at the same time.[8] He says it took no more than about fifteen minutes for him to adapt to typing and mouse clicking while walking. This might seem crazy to some, but it makes sense. Remember, we evolved to be upright and active, not planted in chairs twelve hours per day. Medina envisions a total revolution in the offices and classrooms of the future if society ever catches up to the science. "If you wanted to create an education environment that was directly opposed to what the brain is good at doing," he explains, "you probably would design something like the classroom. If you wanted to create a business environment that was directly opposed to what the brain is good at doing, you probably would design something like a cubicle."[9] I can see the future. No more workers imprisoned and immobile in their cubicles. No more students chained to desks. There will be plenty of standing and lots of movement, and we will be better thinkers as a result. Can't wait.

Gretchen Reynolds, author of *The First 20 Minutes: Surprising Science Reveals How We Can Exercise Better, Train Smarter, Live Longer*, noticed that all the brain scientists she interviewed while writing her book had at least one thing in common: "Every researcher I spoke with on this topic exercises. Some run. Some walk. There are a few bike racers. Tennis is popular, too. But none are sedentary. They know too much."[10] I'm not surprised by that at all. When you understand what science has revealed about who we are, where we come from as a species, and how activity helps our brains and bodies work well, feel well, and last longer, it's difficult to stay in your seat. For the record, I wrote the last several paragraphs while standing.

SLEEP

On a quiet night, I can imagine hearing faint echoes of my mother yelling for me to put down the comic book and get to bed. My brain and I thank you for your diligent parenting, Mom. Everyone knows that sleep matters. But do you appreciate how important it is to your brain? When you stay up late to hang out with friends or watch TV, your brain suffers severely. Shorting it a few hours of sleep may be the rough equivalent of whacking yourself in the head with a croquet mallet. Insufficient sleep can reduce concentration, alertness, and the abilities to learn, problem solve, and remember. And, of particular interest to aspiring skeptics, a tired brain is less efficient at reasoning, assessing claims, and dismantling bad beliefs.

A well-rested brain is an alert brain. A sleepy brain is a goofy brain and, in some cases, a dangerous one. An estimate by the National Highway Traffic Safety Administration puts the number of police-reported car crashes in the United States due to sleepy drivers at 100,000 per year. These accidents account for more than 1,500 deaths, 71,000 injuries, and a financial cost of more than $12 billion.[11] Most responsible people wouldn't think of showing up at work or school drunk or driving under the influence of alcohol. But how many of us have turned up for work or school with a sleep-deprived brain or driven while under the influence of a sleepy brain? One problem with getting enough sleep is that scientists can't even say for sure how much of it we need. There is too much variation and there are too many factors involved to determine a specific amount that is best for everyone. Seven to nine hours seems to work for many, but you might be fine with just six, or you might require no less than ten. Whatever your number is, you need to figure it out and adjust your life accordingly. Probably the simplest way to discover your personal sleep needs is to keep a simple diary for a few weeks. Nothing too complicated, just log how many hours you sleep each night and how you feel during the days. Over time, this might reveal if you are more or less alert and productive based on the number of hours you sleep.

Don't forget to work in brief afternoon naps when you can. Brains love them. Sleeping well and enough is like eating well and staying physically active—you have to pay attention and do the right thing most of the time, otherwise you will be burdened with a substandard brain. If that's not enough to motivate you, maybe fear of pain and death will. Failing to get enough sleep over long periods of time increases your chances of getting a variety of diseases and dying early. Sleep deprivation has been linked to significantly higher risk for heart disease, cancer, diabetes, strokes, and obesity.[12]

What the brain does during sleep is still mysterious in many ways. It certainly doesn't seem to actually rest while we are asleep. According to brain scientist Medina, it's buzzing with activity, probably replaying what we learned and experienced the previous day.[13] It's thought that all this activity is necessary to sort out our memories and file them away for future use. Whatever our brains are doing during sleep, it's definitely important and critical to the quality of the following day. Make sure you remember that there is no ambiguity here, no doubt whatsoever about the devastating effect missing sleep has on mental processes. Moms are right. Get to bed on time!

USE IT OR LOSE IT

Try as I might to maintain some ideal, year-round Zen balance in my life, I never quite achieve it. During the final couple of months while writing a book under the weight of a brutal deadline, for example, it's inevitable that the volume and quality of my physical activity suffers. I still get in an average of thirty minutes of activity most days—have to or my brain won't work right and I feel cranky—but as the deadline nears, the book increasingly consumes my thoughts, time, and energy. Running mileage plummets, and intense, heavy-lifting torture sessions at the gym morph into wimpy high-rep play with light weights. Then, when I finish the book, I switch back to beast mode and resume the eternal and impossible quest for peak strength and endurance. I just can't burn the candle, full flame, at both ends no matter how much I want to. It probably wouldn't be wise to try, so I don't worry about it. I accept this as the rhythm of my life. Rather than feel bad about the loss of my high fitness level, I've come to look forward to the challenge of working to get it back after the book is finished. This cycle that I go through alerted me to the need to keep reading, thinking, and learning, every day, year round. My brain is one "muscle" that I do not want to get flabby and weak. For the brain, there is no off-season, no downtime. Even when I'm not immersed in writing a book, I make sure to continue researching, learning, exploring, wondering, thinking, and doodling. As a result, I'm able to keep my brain functioning at or near the top of its game. I don't want to ever have to get my brain "back in shape" because it's too valuable and necessary to my life every day. I need it in shape and battle-hardened at all times for my overall health and happiness.

Forget snake-oil scams that promise to keep your brain razor-sharp and healthy forever with a pill or potion. Scientists have discovered that what actually works is stimulation in the form of *learning* and *discovery*. The human brain is a kind organic machine that thrives and operates best when it is consistently called on to do its intended job of learning and thinking. It's the kind of engine you want to keep running rather than start and stop.

Everybody knows that baby brains need to explore, experience the world, and learn new things. What most people apparently don't know is that *adult* brains need to do the same things. Grown-up brains also must explore, experience, and learn. If they don't, they whither and weaken. When you stop and think about it, this is one of those things that seem very much like common sense. After all, to be good at something usually requires doing that particular thing. If you want to be good at running, running is required, right? Well, if you want to have a brain that thinks well, then think!

Like muscle and bone, the right amount of good stress in the form of mental work makes the brain strong and capable. And, just like muscle and bone, the lack of it makes the brain weak and frail. There is now some good research, for example, that indicates knowing two or more languages makes the brain better at processing information overall.[14] There is just something about flipping back and forth between languages that keeps a brain sharp and healthy. It even seems to delay for years the onset of Alzheimer's disease in elderly people.[15] According to other research, the act of learning a new skill such as juggling enhances the brain's connections, and connections are the key to thinking well. It doesn't matter if you become very good at juggling or whichever challenging new skill you try. Just making the effort to learn it and practice it is the key to giving your brain a boost.[16] Connective wiring, or "white matter," as neuroscientists call it, is crucial to a healthy brain. The better the network of connections are, the better it works. So, if white matter is good, how do we get more of it? Well, you certainly won't find it in a bottle at a health-food store, and you can't buy any by calling a 1-800 number you see in some late-night infomercial. You have to earn it the old-fashioned way: by learning new things. New challenges for the brain stimulate the growth of white matter, and the positive effect can keep paying off for you late into life.[17] Don't worry if picking up Swahili in your spare time and juggling bowling pins on your lunch hour don't appeal to you. Luckily, there happens to be an entire universe ready and waiting to supply you with an infinite number of things to discover and learn about. Buy a telescope or microscope and start exploring. Learn to sculpt clay. Sketch, learn some new math, pick up a musical instrument and go to work, just make sure you keep challenging your brain in new ways. By the way, another valuable benefit to all this can be an enriched life. The more we have learned and experienced, the more we have lived.

Regular reading, no surprise, is good for you. But not only can this activity lead to a fuller life through knowledge, it also seems to help with your brain development and long-term mental fitness. Scientists have found strong evidence that reading (and writing) can help preserve the "structural integrity" of brains.[18] I'm sure you will agree with me that having structural integrity inside your head sounds like a good idea.

Never underestimate the power of a teacher to influence a student. Nick Wynne, my favorite professor in college, pulled me aside one day and said, "You need to read incessantly." I nodded in agreement. Later, back in my dorm room, I looked up *incessantly* to find out what it meant. I can't say that my reading habits changed immediately, but that one, tiny statement stuck. Looking back, I suspect it was Dr. Wynne's polite way of saying: "Guy, you are in danger of becoming a permanent idiot. Chase fewer girls and read more books." The important thing is that it influenced me. I had loved books since childhood, thanks to many trips to the public library with my mother, but he was right. I needed to up my game. Within a few years, I did become an incessant reader. Today, I don't go anywhere without a book. I keep extra books in my car—just in case I need them. I'm always prepared for those inevitable times when I have to sit in a waiting room or stand in line somewhere. Reading keeps me from being one of those wretched souls at the Department of Motor Vehicles who sit there, impatiently staring at their number, hoping to be called next. Long plane rides may be boring misery for some, but jetliners are wonderful, airborne libraries to me. Down time is never lost time because I read. My motivation to devour more and more books was to learn, but now it turns out that I have probably benefited in ways I never could have imagined years ago. Reading is like taking your brain for a jog or a swim. By scanning the brains of test subjects while they were reading, researchers found that the brain gets a vigorous and unique workout while reading. Scientists discovered "dramatic and unexpected" increases in blood flow to regions of the brain that don't normally have anything to do with reading.[19] Reading, it seems, is valuable well beyond the acquisition of new information. So make sure you read *incessantly*.

Don't spend your whole life inside a book, however. There is too much world out there to miss. And experiencing it is also good for your brain. So go outside and connect with the natural world on a regular basis. Visit a mountain, a desert, a forest, a pond, a beach. We think of ourselves as thoroughly civilized and sanitized. But we are still creatures of the wilderness. Dirt, plants, trees, wild animals, and sky matter to us. And these things make our brains feel good. One of the best things you can do for your brain on a regular basis is join exercise, thinking, and nature into one outing. The closest thing to a miracle tonic for the brain might be something like a vigorous walk/run on a previously unexplored mountain trail or along a new stretch of beach, during which one notices and photographs an unusual plant or insect to research after returning home. That would be a good day for any brain.

Traveling to new countries is also wonderful brain stimulation. Don't be intimidated by cost or safety concerns. I've been all over the world, to every continent except Antarctica, and I did most of it alone and with little money. If you do your research and don't mind giving up massages and lobster tails, it is possible to experience much of your planet safely and economically. So if your international travel extends no farther than a few laps around the World Showcase at Epcot, it's time to get off the porch. Few things stimulate your brain and make you feel alive as taking in the smells, tastes, sounds, and sights of faraway lands.

Whether at home or abroad, make sure to take full advantage of those very special buildings where art, history, culture, and science are preserved and displayed. I'm not ashamed to admit it, museums are my cathedrals. I don't actually worship the artifacts within or anything weird like that, but the better ones really do stir up an abundance of awe and reverence in me. My heart rate soars in joyous anticipation when I walk up the steps and toward the entrance of a great museum. (I'm not joking.) All that excitement, learning, and creative thinking that good museums spark in me *has* to be good for my brain.

This has been a long love affair for me. When I was very young, my grandfather took me to museums, and I still remember seeing my first shrunken head in a glass case. I loved seeing dinosaur fossils two feet in front of my face and being able to touch World War II airplanes. I have no idea if my grandfather took me as part of some noble attempt to turn his goofy grandkid into an intellectual or if he just got stuck with me on those days and had no choice in the matter. But it doesn't matter because I fed off his enthusiasm for museums and became hooked for life. Being surrounded by so much mental stimulation in one place is fun and, more important, very healthy for the thinking machine. Whenever I visit a city or town, one of the first things I do is head for the nearest museum. Try it.

If you ever have doubts about the joy and benefits of being eager to learn, just look at young kids. Most of them are profoundly inspiring. They know how to use their brains and don't hesitate to do so. They want to learn. They *have to learn*. Kids ask questions. They explore, experiment, touch, and imagine. I used to teach science and history to elementary and middle-school students, and I will never forget how curious and excited they were about discovering new things. Before and after lectures, many of them would run to me and blurt out incoherent questions. They were so eager to learn something new that they couldn't even slow their brains down to formulate their questions. Those kids inspired me in a very lasting way. I don't care how long I live, I'll never be so stupid as to imagine that I've heard it all and seen it all. I'll never be so dead inside that I wouldn't want to know what might be living under that rock up ahead or what might be waiting around the bend. And my brain is better off for this attitude.

Become a lifelong student and never stop exploring. Dreaming and thinking must never cease. More thinking makes you a better thinker. The more you learn, the more you want to learn. Do this and you are sure to create a wonderful circle of reinforcing strength that will serve you well for the rest of your life.

GOOD THINKING!

- Brains do so much for us, yet most of us neglect them and do not care for them as well as we should.
- Good nutrition is vital for the proper care and maintenance of your brain. Try to do the basic and obvious good things, such as eating lots of leafy, green vegetables and keeping the total amount of food consumed on a daily basis reasonable. A poorly

fed brain cannot think well. A brain that is trapped inside a poorly fed body cannot be optimally healthy.

- To work well and stay healthy, brains need to be inside of active bodies. It is crucial that you do some form of physical activity for at least twenty or thirty minutes, six days per week.

- Brains need sleep. There is still mystery about what brains do during sleep, but whatever it is, it's important. A sleepy brain is a substandard brain.

- Staying mentally active throughout life is key to keeping the brain functional and healthy. Learning new skills creates new connections in the brain, which are linked to overall health.

Chapter 5

SO LITTLE TO LOSE AND A UNIVERSE TO GAIN

There are few basic questions that probably stop many people just short of transforming themselves into good skeptics with a passion for learning and thinking: *What will I lose? Is it worth it? Will I be okay if I give up some of my beliefs?* These are understandable concerns. I don't dismiss them or take them lightly. Many people are just not sure about shining the hot light of science and reason on their beliefs because they worry that doing so might result in them having to make sacrifices that they just don't want to make.

Many people love their beliefs and claim to need them for the comfort they provide. Others view their beliefs as an invaluable and irreplaceable source of thrills that brighten and energize what would otherwise be a boring and dim existence. And some say it is beliefs that give meaning to their lives. Often it is the same person who says all of these things. They cling to one or more dubious beliefs because they rely on them for comfort, excitement, and meaning—the whole shebang. People who feel this way don't just want beliefs. By their own admission, they *need* beliefs. But is this true? Do they really depend on them to the degree that they think?

I could be wrong, but I don't think that most people need to accept claims that probably are not true in order to have at least a fair chance of finding or achieving acceptable levels of comfort, excitement, and meaning in their lives. Maybe the typical believer in psychic readings or astrology really would be reduced to the psychological equivalent of a warm blob of twitching jelly if she or he realized one day that her or his beliefs were unlikely to be true. But I doubt it. I know too many people who let go of their beliefs and have been fine ever since. In my opinion, virtually all people—whether or not they know it—are smart enough and strong enough to cope with the difficulties and uncertainties of life without needing to wrap themselves in fantastic claims. I only have to look at the example of my own life. I'm not so lucky that I escape facing many of the same challenges and disappointments other people do, yet I have managed to persevere so far without leaning on irrational beliefs. In fact, I suspect that I've coped with the stresses of life *better* than I might have otherwise because my head is less cluttered with bad ideas and distracting delusions. This allows me to focus on the problem at hand and, if necessary, rely on real people who can help or comfort me. But that's just me. Far more significant is the example set by the many millions of people around the world who don't believe in most or any of the usual unlikely-to-be-true claims.

So how do they do it? How are they making it, all alone so to speak? By living in the real world and dealing with life the best they can, that's how. These nonbelievers aren't invulnerable superheroes. They aren't emotionless Vulcans. They suffer. They fail. They endure losses. They cry. But somehow they live and succeed without pretending to know things they don't and without constantly explaining things that currently have no explanation. In short, they act like mature and brave human beings. The existence of good skeptics who thrive despite having no belief in ghosts, magic, miracles, and so on may not disprove the notion that people need to believe, but they certainly cast doubt upon it.

Yes, it is disturbing that about a quarter of the American adult population believes in astrology and 37 percent believes in haunted houses.[1] But let's not overlook the reverse side of such statistics. Somehow, *75 percent* of adults manage to get up in the morning, dress themselves, and face the day even though they *don't believe* in astrology; and 63 percent of adults seem to find life interesting enough *without believing* in haunted houses. Overall, about three out of four Americans believe in at least one paranormal claim—but that leaves a quarter of all Americans who do not.[2]

If belief is critical to the good life, then what are we to make of people who don't believe? Are they all miserable misfits and dysfunctional dolts? That's obviously not true. So are they a superior race of people with advanced brains who have evolved beyond irrational belief? Nope, not true either. It's safe to assume that most people who are living a relatively belief-free life are a mix of regular people much like everyone else. They may have made the decision to think critically and do their best to live in the real world, but apart from that, they are more alike than different compared to the rest of the population. Even religion, so often hailed as *the* prerequisite to a good and meaningful life, somehow isn't necessary for the more than one billion people on Earth right now who describe themselves as nonreligious or unaffiliated with any religion.[3] Between 500 million and 700 million of these people go one step further and describe themselves as atheist or agnostic.[4] I don't doubt that unproven beliefs can provide comfort, excitement, inspiration, and meaning for people. I just think that these things can be found in known reality as well, as a large portion of the world's population proves every day.

TOO BIG TO FAIL?

A key challenge for the good skeptic is the common mind-set that says some beliefs are too special to question. To me it's obvious that nothing should be off-limits to sincere and sensible inquiry. If we can't honestly analyze and reconsider certain "important" beliefs, then how can anyone ever really know if those beliefs are valid? It makes no sense to say that something is too serious to treat seriously. Scrutiny should *rise* with importance, not go down. But many disagree. They feel that some beliefs are simply too big to fail and can't be challenged because if they collapse, too many people will go down with them. I don't buy this, of course, so I don't refrain from encouraging skeptical thinking when it comes to any particular extraordinary or important claims.

My motivations are all good here. I want only to be constructive and help others. I look around and see a world that is deeply wounded and severely burdened by irrational beliefs. This drives me to act. I'm no saint or superhero, but I'm willing to at least try to improve the world. It seems obvious to me that we can do much better than this. Every day, I read news reports of horrors and heartbreak tied to one irrational belief or another. Skepticism is the prevention and the cure. One incessantly hears about money troubles these days and somehow the world just can't scrape up enough cash to feed and vaccinate every child. Yet humankind currently spends hundreds of billions of dollars per year on fortunetellers, medical quackery, and other nonsense. We also seem to be constantly short of time and always tired, but somehow we are able to find countless hours and come up with enough energy to worry about things that probably aren't real. Why are we crippling ourselves in this way? It's the twenty-first century. Skeptical thinking should be more popular by now. After all, it's not a state secret that science works. The fact that the scientific process does an excellent job of separating truth from fiction is not forbidden or inaccessible knowledge. Virtually everybody today relies on science to some degree—even the Amish. So why not rely more on scientific thinking?

One doesn't have to make very much effort to find good reasons to be a critical thinker these days. Thanks to the current state of television news, it's easy to come by examples of how failing to think before believing can lead to problems. The human misfortune and misery that results from weak skepticism is an endless source of fodder for TV news. Just turn the television on and marvel at the madness: *On advice from his spiritual guide, man in Fresno eats four pounds of carpet lint and then drinks seven pints of Komodo dragon saliva! Full report from the intensive care ward at 11:00 tonight, plus NFL scores and weather!* Our modern world is a nonstop parade of people being punished for their failure to think. Why don't more people recognize this and react sensibly by deciding to think for themselves and question everything? Are they afraid of something? Are they waiting for something?

Being a good skeptic is not scary. What's scary is letting other people think for you and then hoping that it all works out well. It's not difficult to be a good skeptic and accept reality. What's difficult is figuring out how to pay the rent after giving half your paycheck to a faith healer. Difficult is hearing your doctor tell you that it's too late. She could have saved your life with medical science, but now you are going to die because you spent the last year harmonizing your spirit energy with the universe and consuming fungus-infused shark pills sold to you by some crook with a smile who promised they would cure you.

Things have changed since we first began this routine of making up answers to mysteries and accepting claims without evidence. We invented the microscope and telescope more than four centuries ago and have seen many things since. We now know how to discover the real and expose the fake using science. History is clear: bad thinking often leads to very bad outcomes. Yet today many smart and good people remain enslaved to irrational beliefs. Some hate and kill in the name of gods. Many expect magic to heal them when ill. Others see aliens, ghosts, and conspiracies lurking in every shadow. We rode on the back of science all the way to the Moon, yet millions of people today reject science and insist that stars and planets hold magical, predictive powers over our lives. What is going on? Who are we? What's wrong with us? One scientist summarizes humankind as having "paleolithic emotions; medieval institutions; and god-like technology."[5] That is more than a description; it's a warning, too. Clearly this is not a stable mixture for us as we barrel ahead into an increasingly high-tech and complex future. By 2050, there likely will be more than nine billion people sharing this same planet. It is difficult to imagine how we will avoid a pathetic series of self-inflicted catastrophes of increasing magnitude if we do not recognize the need and find the will to finally become less a believing species and more a thinking species.

My reaction to this critical challenge is to encourage people to become good skeptics. More thinking by more people ought to lead to better results for us. Maybe it won't, but it seems to me like it is well worth trying. More skepticism means less irrational belief, which probably means a better and safer world. Skepticism can't solve every problem, of course. But it certainly will help a lot. I hope people don't view this as a fight between skeptics and believers. I prefer to think that we are all in this together. Living on a fantasy-prone planet ought to be a common concern because we all pay a price for it, regardless of who we are or where we are. Most believers probably would agree that there are far too many irrational beliefs out there today. They may not recognize a problem with their own belief, but they certainly recognize the drag or danger in others'. It's a start. There is common ground here. For example, I am sure that a significant number of people, believers and skeptics alike, agree that Christians torturing and killing "child witches" in Africa and Hindus abusing and killing "sorcerers" in India is not only criminal and immoral but also profoundly senseless. If so, then promoting science and reason is something we can support together. Similarly, millions of women today are openly treated as the property of men. They are oppressed, beaten, and sometimes killed by men who believe their actions are sanctioned, if not required, by belief systems that any good skeptic can challenge easily. These acts of violence constitute yet more common ground, one more reason for both believers and skeptics to promote rational thinking.

CONSTRUCTIVE OPTIMISM

Most good skeptics are not at war with typical well-meaning believers. I'm certainly not. I view my fight as being *against* bad beliefs and *for* the good people who believe them. In a similar vein, I think malaria is bad, but I certainly don't have any animosity for its victims. What sense would that make? The kind of skepticism I'm promoting is constructive and optimistic. In my own small way, I'm trying to be *constructive* by helping to improve the world for all of us. Because I'm *optimistic*, I think our species has the potential to become more rational. So when people indicate that they are not comfortable asking questions and demanding evidence because it might cost them a few of their cherished beliefs, I appreciate their concern but don't hesitate to explain what unproven beliefs have done and continue to do our world. I also mention what the skeptical lifestyle offers. It's the greatest weight-reduction plan of all time, one that actually works. It trims away the mental fat and turns people into lean thinking machines capable of running strong and long rather than stumbling passively through a possibly diminished life. I don't disregard believers' fear of loss because I know it is probably sincere. Giving up beliefs that probably aren't true may not be that big of a deal in my view, but the perception some have that it is a big deal makes the concern important enough. Yes, something will be lost when skeptical thinking is allowed to run its course. The loss is not so great relatively, however, because of what one stands to gain: *everything*.

Why are many people reluctant or uninterested in becoming good skeptics? Is it because they haven't heard enough about the wisdom of skepticism? That's one problem, of course, but there is an even bigger one. Many people are afraid that they will lose social footing if they embrace skepticism. If their new way of thinking corrodes or dissolves the warm blanket of superstition they have wrapped around themselves, not only could they end up out in the cold but also they might find themselves suddenly alone, without friends, family, and community support. So on top of a conviction that says their unproven and unlikely-to-be-true beliefs give them comfort, strength, and meaning is the fear that abandoning those beliefs could leave them socially isolated. This is not a trivial or unreasonable concern. Beliefs do serve as a social glue or platform for building and maintaining relationships of all kinds. The best response to this, however, is to point out the obvious: There are many ways in which people come together and enjoy meaningful connections that have nothing to do with unproven beliefs. Again, remember to consider the example of the many people who make their way in life without silencing their skeptical minds. They're not all hopeless hermits doomed to dwell on the cold, dark fringes of society. They have families. They have friends. However, no one should deny that it can sometimes be very difficult to fit in and be accepted when you don't believe as others do. Good skeptics have to realize that every moment doesn't need to be a clash of philosophies and a debate over evidence for this or that. The "live and let live" attitude works well at the party, on the street, and in the office most of the time. Included among my friends, for example, are people who seem to believe every kooky and bizarre claim ever to be inflicted on an innocent brain. Although I may worry about them at times, I still think of them as valid human beings in every way, of course. Their friendships add value to my life. I may raise questions about their beliefs from time to time, but I don't feel the need to vigorously challenge them on a daily basis. I can offer ideas and encourage them to think, but ultimately it's up to them.

It also has been one of the joys of my life to discover that so many believers can and do accept me for who I am. This is not to say that I have not been rejected, had my feelings hurt, and been mistreated by some because of my skeptical ways. It's never going to be a consistently smooth ride for the good skeptic, not on this fantasy-prone planet. But does anyone anywhere enjoy a smooth ride all the time? Don't we all hit the occasional bump in the road? The trick is to just keep going. Skeptics aren't alone in this. Even believers deal with acceptance and rejection issues. Does the New Age connoisseur of magic crystals find herself welcome among Pentecostals? Probably not. Does the Christian feel safe at a Taliban prayer gathering? Definitely not. We all have moments in life that bring tension and the possibility of conflict or rejection.

Fear of losing friends or complicating family relationships is not a sufficient reason to stop from becoming a good skeptic. No doubt other people matter. We are social creatures in the extreme. But surrendering or sacrificing one's mind out of fear and insecurity is not the best way to maintain membership in the human club. Your brain is the one thing that makes being human special. Use it, become a good skeptic, and figure out the social stuff as you go along. If you happen to be in extremely complicated circumstances, say you are stuck on an island with people who think dragons are coming to save the world and these people might kill you if you question their claims, then use common sense and keep quiet. If you are at a place in life where you depend on believers for food, shelter, or security, then don't bother mentioning that you have analyzed their ideas and found them to be ridiculous. Just keep quiet and smile whenever they talk about dragons. Meanwhile, take pride in your independent thought, be happy you aren't deluded like them, and work on finding a new place to live.

THRILL SEEKERS

When many people tell me about one of their beliefs, I see excitement in their eyes and hear enthusiasm in their voices. While I do understand the buzz behind belief in guardian angels, the Roswell UFO story, or a precise and imminent doomsday date, I do not see any kind of irreplaceable value in these kinds of claims specifically. We can find plenty of thrills in other places, too. Furthermore, isn't a joy or thrill that comes from a mistake, delusion, or lie a kind of counterfeit high anyway? Sure, it might feel good—but for how long and at what cost? If some practical joker gave me a fake lottery ticket with a winning number on it, I might feel really, really good—right up to the moment when I tried to cash it in. My previous joy wouldn't be enough to make the disappointment and anger worth it. Likewise, getting drunk or high might feel good in the moment, but spending every day of one's life that way might result in self-destruction and a wasted life.

I'm not denying that some beliefs can add spice to a life. What I don't accept is that these beliefs are the only way or the best way to elude boredom. Yes, believing in amazing and weird things can be fun and exciting. However, embracing beliefs that fall short on evidence and logical arguments because one gets a kick out of them seems like cheating to me. It's turning off your brain in order to grab a cheap high. It's like closing your eyes and spinning around in circles on your front lawn while imagining that you are on the roller coaster. Why the charade? Why not just open your eyes and go find a real roller coaster?

I often wonder how much thought believers give to what reality offers them. Why worry about invisible mystical forces that science can't detect, measure, or confirm when there are real things like nebulae and black holes out there? Why chase mythical monsters when all around us there are more bizarre creatures that exist than we can ever know in a lifetime? The real universe is not a bitter pill to be choked down. It's the ultimate theme park and your human brain is the golden ticket. This is a place of infinite excitement and mystery. Depending on where you are and where you look, it offers many comforts and many dangers.

When I was a young boy in Florida, I remember finding a Portuguese man o' war that had washed up on a lonely strip of coastline I was exploring. I was blown away by the bizarre creature. It was so alien in appearance, so unlike anything I had ever seen before. It was at once monstrous and beautiful. The longer I looked at it, the less it seemed to belong on this planet. Its purple-tinted, gas-filled, transparent bubble enabled it to float at sea. It was shaped in a way that made it an effective sail for traveling many miles in the ocean. I was amazed by the creepy and dangerous purple tentacles. I tried to imagine how much it would hurt if I touched one briefly. I chose not to find out. I already loved science fiction in those days, so extraterrestrial life was one of my many interests, and this thing pushed my buttons, to say the least. I knew it was an Earth life-form, of course, but its freaky structure led me to think more deeply about what life might be like on other worlds and how there is so much variety right here. I disagree with those who suggest that extraterrestrial life, if it exists and if we discover it, would be so exotic and unexpected in form that it would shock everyone. I doubt this because our own planet has such extreme diversity that if we ever did find anything, chances are a zoologist, microbiologist, or marine biologist somewhere would immediately say, "Hey that looks just like . . . " The alien physiology, how it functions, likely would be differently from anything we know, of course, but chances are its appearance would be similar to *something* here on Earth. That's just how wonderfully weird and varied Earth life is. It's almost like we have an entire universe of life right here with us. All we need to do is notice it.

Fast-forward to adulthood, and I'm giving a science lecture to a class of very bright students in the Caribbean. We're talking about ocean life and I'm showing beautiful color photographs of amazing animals only recently discovered in extremely deep waters. I mention to the kids that many of the creatures seem like aliens, as if they don't belong on Earth. But then I wonder aloud if life on a faraway planet would look like anything on Earth. Maybe they are so different that we don't have anything of a similar shape and structure on our world. Maybe they are intelligent clouds of organic fog. Maybe they are liquid and flow together to create one big conscious ocean. Maybe their "bodies" are a network of atoms spread apart over vast distances in space. Or maybe they don't exist at all. Maybe we are alone in the universe. Maybe we are the first life ever. Or maybe we are the last. I have no idea, but I love asking the questions and imagining possible answers. That was an exciting class for me and for the students that day—even though we never once relied on irrational beliefs or felt the need to accept any unsupported claims.

Back on that Florida coastline so many years ago, I probably stared at the Portuguese man o' war for at least an hour. In kid time, that's like ten hours. I resisted my little-boy urge to pop the bubble with a stick and instead pushed it back out to sea. I couldn't tell if it was alive or dead, but if it was, I was determined to save it. Anything that strange and that cool deserves to live, I figured. From that day on, the man o' war would always be one of my favorite marine animals and easily my number-one jellyfish—only it's no jellyfish.

I later learned that the creature I had encountered was even stranger than I thought. It turns out that the Portuguese man o' war is not a jellyfish at all. It's not even *one* animal. Each man o' war is a *colony* of many individual creatures. They team up to create a floating, carnivorous fortress that sails the seas in search of fish to paralyze and eat. I also learned along the way that these amazing animals can release the gas from their bubble to submerge like a submarine when attacked and that sometimes their tentacles grow to more than one hundred feet in length. There is a lesson in this story somewhere: Thinking like a scientist is the gift that keeps on giving. It's a never-ending process of discovery. New information is always coming in, or will if your brain is open to it. I am no less curious today than I was as a child. I don't cringe when I find out I was wrong about something; I celebrate the upgrade to my brain.

This willingness to adapt and change for the better is important to being a good skeptic. As a lifelong fan of science, I have no choice but to be humble. I can't imagine being any other way. I try to explore and learn new things every day. And the more I learn, the more I sense how deep and wide my ignorance runs. This is not a bad thing. My awareness of it makes life more exciting. I love knowing that I'll never run out of things to learn and experience. This is one of the reasons I find it nearly impossible to get excited about extraordinary claims about events, abilities, or monsters that don't seem reasonable and lack good evidence. I know about too many wonders that actually do exist or are more likely to exist than to get all worked up over things that almost certainly do not.

WONDERS LARGE AND SMALL

Our planet's microbes (microscopic life-forms) remain mostly unknown to us. This is a strange state of affairs given their abundance and importance. It's like there is an entire parallel universe of life right here with us and yet we understand relatively little about it. Despite our limited knowledge, however, we have managed to figure out that they pretty much run the planet. Without microbes, civilization would collapse in a month. Our own bodies can't even function without them. Not only are they around us, on us, and in us, but also microbes live in the muck at the bottom of the deepest ocean, miles below the surface of land in near-solid rock, and even up in the clouds above us. If you are attracted to mystery and monsters, then give microbes a chance because they've got it all. I can hardly look down at a patch of common dirt without hearing the voice of Edward O. Wilson, a scientist who specializes in ants but has a fondness for life even smaller. He admits that if he could take another lap in life, he would make a career in microbes:

> Ten billion bacteria live in a gram of ordinary soil, a mere pinch between the thumb and forefinger. They represent thousands of species, almost none of them known to science. Into that world I would go with the aid of modern microscopy and molecular analysis. I would cut my way through clonal forests sprawled across grains of sand, travel in an imagined submarine through drops of water proportionately the size of lakes, and track predators and prey in order to discover new life ways and alien food webs. All this, I need venture no farther than ten paces outside my laboratory.[6]

Those who choose paranormal promises over scientific realities have no idea what they are missing. For example, I recently learned that *Desulforudis audaxviator*, a microscopic bacterium, has been found living more than half a mile down in the ground near Death Valley, California. This is believed to be the same species that was found a few years earlier living more than two miles below the Earth's surface in South Africa. For hundreds of millions of years, these tough guys have lived in a very hot (temperatures up to 140°F/60°C), totally sunless environment, with no oxygen and no life or organic material to keep them company. Don't miss that last part. It lives *alone*. This is the only known life on Earth to date that lives in complete isolation. So how does it survive? What does it eat? It relies on the by-products of radioactive decay in the rocks around it. But the coolest thing of all about *Desulforudis audaxviator* is this: Scientists think some members of this species take wild rides up to the surface via natural water springs where wind blows some of them thousands of feet up in the air. After traveling halfway around the world, they ride raindrops back to Earth and then work their way back down to where they like it best, dark and deep.[7] I don't know how you feel about it, but I *love* sharing a planet with freaks like *Desulforudis audaxviator*. By the way, lonely as it may be for them miles down in rock, they certainly are not alone during their journeys up in the sky. It turns out that the air above us may be yet another huge and important ecosystem, one totally overlooked until very recently. For example, scientists have taken air samples twelve miles up and recovered more than 2,100 different species of microbes. These particular bugs they found were traveling from Asia to North America in the upper troposphere. It's like they have their own highways up there. Some researchers think that these microbes are not just dormant hitchhikers but are active while in flight, possibly mating, for example. (Twelve-mile-high club?) It's even possible that this microbial high-altitude highway has significant influence on the world's weather.[8]

In addition to land and sky, the microbes seem to rule the oceans as well. The next time you are walking the sands of your favorite beach, glance out to sea and think about what is out there. But forget fish, plankton, and dolphins, and the usual cast of characters. Those are mere bit players in this world. Even plankton, for all their importance, have been upstaged. It's the viruses that run things. They infect *everything* in the ocean, from the smallest bacterium to the largest whale.

It is now clear that the ocean is pretty much one big viral infection. Marine viruses are the most abundant form of "life" on all the Earth. (Viruses are so weird that technically they don't qualify as life. But they're close enough.) A typical liter (0.26 gallons) of ocean water contains more than one hundred billion viruses! That's about the same number of all people who have ever lived throughout the entire existence of humans. A kilogram of mud or sediment from the seafloor can easily contain a million or more *different species* of viruses. Most viruses are much smaller than even bacteria, so they weigh very little, of course. But they are so prolific that their collective weight adds up quickly. For example, all the viruses in the ocean together would weigh as much as the combined weight of 75 million blue whales. Even more mind-boggling, if all marine viruses were placed end to end, they would stretch out into space farther than the next sixty galaxies![9]

Okay, gaudy numbers and dramatic analogies aside, what's the big deal about a bunch of creepy little parasites that float around, infecting fish, plankton, and everything else in the ocean? They matter because they seem to be nothing less than the framework of the entire ocean. And as the ocean goes, so goes life on land. Viruses may be invisible to our naked eyes, but they shape and effectively control the living ocean that we can see, as well as the rest of the Earth's biosphere. It's a common human fault to imagine that we are somehow above or disconnected from the rest of nature. The reality, however, is that we are far more dependent upon worms, bacteria, and viruses than we are on cars, computers, and credit cards. Given the importance of viruses to the way the world is, shouldn't people devote some time to understanding a bit about them? But no, for millions of people, ghost hunting, astrological research, and other such pursuits beckon, and there is only so much time in a day.

Most viruses pose no direct threat to us, by the way. But that doesn't mean they aren't ruining somebody's day. They have their chosen targets, and that's just what viruses do. They can't reproduce without hijacking a cell and turning it into their own personal virus factory, so their mission always is to invade and conquer. Marine viruses kill about *half* of all marine bacteria every day. Overall, they kill about *20 percent* of all ocean life every day. This is amazing, especially when you consider what would happen if all those trillions and trillions of bacteria and other creatures were not being snuffed out by viruses on a daily basis. The food chain we rely on would be very different without viral population control—possibly not so well suited to our needs. For example, scientists have known for many years that the microbial ocean plays a major role in the Earth's atmospheric makeup—and marine viruses have perhaps the biggest say in all of that. So, if you appreciate having food to eat and air to breathe, be sure to thank a virus.

If the virosphere still feels distant and disconnected, consider this: *You* are a virus, at least that's what your genetic code says. Like it or not, viral footprints are all over humanity. This might not go over well with people who still can't bring themselves to invite bright bonobos to the family picnic, but viruses deserve a place at the table too. They have been around for billions of years, and that means they have coexisted closely with plants and animals, including us, so close that we are now part *them*. Our long and intimate evolutionary dance with viruses can't be denied. It happened. It's happening. But don't worry; this is nothing to be ashamed of. We can take pride in having such prominent and influential kin.

As is probably clear by now, I'm so fascinated and excited by this branch of science that it's difficult for me to shut up about it. As a result of living in the same house with me, for example, my otherwise-normal twelve-year-old daughter knows far more about viruses, bacteria, and mites than most college students. What I don't understand is why everyone isn't excited about this stuff. Science continues to pull back the curtain on a breathtaking and important show that's underway. The more we learn, the more it becomes apparent that this is the microbes' world and we're just living in it. For example, "you" are made up of 10 trillion or so *human* cells but this is nowhere near the complete picture. There are about 100 trillion *other* living cells and various microscopic critters living on you and inside of you right now as well. Some one billion bacteria live on every square centimeter of your skin's surface. It's true; it doesn't matter how many times per day you shower or use hand sanitizer. You are less "you" than you are "other" life. You are outnumbered in your own skin.

Right now, there may be herds of tiny *Demodex* mites grazing on your face like buffalo on the plains of nineteenth-century North America. Closely related to spiders, these microscopic beasts look like a medieval mace, the stick weapon with spikes on one end. Regardless of how clean you are or what you may be up to at the moment, these creepy critters are busy taking care of their own business. They mate, right there on your face, for example, and guess where they go with the fertilized eggs for safekeeping? Down into your pores—where else? They prefer your eyebrow regions, and it's believed that these mites feast on the oily secretions of your skin. Eating leads to pooing, of course, so how does that work? I have good news and bad news. The good news is that they never poo on you during their entire lifetime because the *Demodex* has no anus. The bad news, however, is that when they die, their bodies break apart to release a lifetime of poo on your face.[10] See? Reality is never boring.

Here's a trick question you can irritate your friends with: How many visitors from Earth have been to the Moon? The likely answers you will get are: (1) *Thirty or so because the astronauts took some frogs, worms, and bugs for experiments in space* [Wrong, Apollo astronauts conducted no animal experiments on the Moon.]; (2) *None, because the Moon landings were a hoax* [Maybe you should find some new friends.]; (3) *Twelve, the number of Apollo astronauts who landed on the Moon* [Respectable answer, but also very wrong.]. The right answer, of course, is *many, many trillions*. Only twelve astronauts may have walked on the Moon, but every one of them was host to a staggering assortment of microbes at the time, so each landing was really more like a mass invasion.

Yes, you are a bipedal ecosystem, a kind of rainforest or coral reef with legs. Wherever you go, remember that you carry with you a vast and diverse collection of wildlife. You are a wilderness of known and unknown bacteria, viruses, and arthropods. And I won't even bother going into detail about the fungal forests that cover large tracts of your body's surface. Cleary you needn't ever feel alone—because you're not. I encourage you to think about these things when someone suggests to you that all the exciting and cool action is happening over on the supernatural/paranormal side of the tracks. No, reality holds its own just fine.

READY FOR THE BIG TIME

Moving on from the small stuff, you might also consider yourself fortunate to have been born into a universe that is absurdly large and home to so many very large objects. Our universe is so big that it's difficult, if not impossible, to grasp it. For example, even if you had a spaceship capable of traveling at the speed of light, it still would take you more than *150 billion years* to make it from one side of the known universe to the other. Better pack a lunch for that trip.

The Sun is gigantic, right? After all, our Earth shrinks to a tiny spec in comparison. But astronomers know of stars that are much larger than ours. One of them, NML Cygni, is about 1,650 times larger than the Sun! Our planet may feel big to us, but it seems microscopic in comparison to that.

We are all fortunate to live in a time when science has advanced far enough to have built a magnificent portal from which to watch the show. A life without magic certainly need not be dull in this neighborhood. It's not like we don't have anything to occupy us. There is always something to marvel at, something to learn about, and more to discover. There is so much, in fact, that it's absolutely impossible to keep up. Just in the time you spend reading this page, for example, a few hundred thousand new stars will be born. By the way, if your life ever seems too slow, just remember that the Earth is spinning at the equator at a rate of about a thousand miles per hour. We are also flying through space around the Sun at speeds of more than 65,000 miles per hour. Pause and think about this. Right now, you are flying through space faster than a bullet, at speeds the fastest jets do not remotely approach. But we're still slowpokes in a sense because if we rounded up volunteers and put them in the fastest spaceship our current technology could produce, it would take them many thousands of years just to reach the nearest star in our own galaxy.

Although the open space between celestial bodies is mind-numbingly vast, the universe definitely is not empty. Our galaxy alone contains from 200 to 400 billion stars. And the Milky Way is just one system of stars among maybe 200 hundred *billion* or more others in the universe. Some of these galaxies may have as many as a *trillion* stars in them. To save you the trouble, I did the math: *300 billion stars* × *200 billion* galaxies = *More stars, planets, and moons than you or I could ever comprehend.*

Time can be exciting to think about, too. Thanks to the scientific process, we know that the universe is about 13.7 billion years old and the Earth is around 4.5 billion years old. Life here is more than three billion years old at least. We modern humans joined the party only a tiny fraction of a second ago, relatively speaking. Sometimes we may feel like a year is forever, and yet a single year is virtually nonexistent, lost within billions. On my bedside table, I keep a beautiful trilobite fossil, and not a day goes by that I don't glance at it. I sometimes think about what a privilege it is just to be able to know of something that lived hundreds of millions of years ago. Our lives may be only tiny flashes of existence in the universe, but thanks to our brainpower, we at least have the opportunity to contemplate our place, to appreciate the time we do have, and to find some joy in it.

I can understand why some people might think about the span of life on Earth or reflect on the size of the universe and feel uncomfortably small or insignificant—but they shouldn't. A human life might not seem like much in the shadow of so much space and time, but that is only one way of looking at it. Another way would be to get excited and feel fortunate to belong to a species that is not only intelligent enough to partially figure this stuff out but to also appreciate the fact that we are a part of it all. For example, the scientific process has revealed the age and origin of the atoms that we are made of. All the hydrogen atoms in you at this moment are almost as old as the universe itself. They were created shortly after the Big Bang, some13.7 billion years ago.[11] Your other atoms were forged in stars and are younger but are still billions of years old. You are ancient. You are part of something larger than yourself. No one should think about the universe and shrink. When I look up at night, I feel a thousand feet tall. After all, we are literally a part of this magnificent and mysterious spectacle. The universe is us, and we are the universe, right down to every single atom in our bodies. I first heard or read about this link to the stars long ago as a child, but it took some years for it to sink in. As it did, however, my perspective changed. It was one more reason to feel free of the borders and walls that others had constructed around me. I became a citizen not just of the Earth but of the entire universe. Today I feel that any other outlook would dull and diminish my life.

These are exciting times of discovery and change. And here you are, lucky enough to have a front-row seat to it. Right now you are—or should be—witnessing the continual unveiling of a profoundly important and fascinating reality of outer and inner space that was mostly invisible and unknown to 99.999 percent of our ancestors. I recommend you go all in and dedicate yourself to learning as much as you can about the real universe, the real Earth, and the real you. If others insist, let them continue to spend their time listening to psychics who charm them and tell them mundane things about themselves that they already knew. Let them buy and wear "ion-infused" sports bracelets to enhance their balance as they walk off mental cliffs. Do the decent thing and help them if you can, of course, but don't let them hold you back. You have important things to do with your time. You have universes to explore, both near and far.

Science is exciting because it leads us to real wonders. So much is happening right now and so much more is coming. Twenty or thirty years ago, it seemed reasonable to assume there were planets beyond our own solar system, but no one could say for sure that there were. Now, however, thanks to improved search abilities, scientists are finding new ones almost every day. Now we can say with unprecedented confidence that our galaxy likely contains hundreds of billions of planets and moons. As ignorant as we still may be, what we do know via science is enough to excite and interest me for a thousand lifetimes and more. How about you?

Don't neglect human history and prehistory by focusing too much on all the exciting action beyond humanity. Like science, there is just too much history to know everything, but no thinkers should be walking around with massive gaps in their knowledge of our past. Make sure you aren't one those people who never heard of the Australopithecines or Louis Leakey. Find out the basics, at least, about what archaeology has revealed regarding the rise of civilizations. I also advise not getting too caught up in *your people's* history or *your country's* history as many do. Learn that, of course, but do make sure to take in the big picture. The truth is that we are a young and closely related species, so *all* history is *your* history. None of it should be dismissed as irrelevant or unimportant. Regardless of who you are and whatever labels you had pasted on you at birth, you ought to feel sadness for the bad parts of world history and take pride in the good parts because all of it is about your family.

Even if you made As in middle- and high-school social studies classes, you still might have some work to do. Knowing which countries won the big wars, who sat on the thrones, and who won the elections is not enough. It's important, and fun, to learn about what the regular people were up to, too. Their lives were not recorded as often or in as much detail as the lives of the rich and powerful. This is why archaeology is so important. By finding the art, tools, toys, and garbage of everyday people long ago, archaeologists can construct very good snapshots of how they lived and what they did with their time. Too many people still think that history is a stale compilation of cold facts and dates. It definitely is not. History is the greatest story of all. History and prehistory are about a scruffy yet clever bunch of primates who were unafraid to think and as a result rode their brains all the way to spectacular highs and shocking lows. Please, know your story.

THE MEANING OF LIFE

I'm reluctant to tackle the topic of life's meaning because it is so difficult to address in a sensible manner that is of use to anyone. So many words and hours already have been sacrificed in vain on this one. It's so personal and subjective that what people say about it usually ends up having about as much weight and substance as the gas inside a Portuguese man o' war's flotation bladder. However, I think that good skeptics who wish to engage believers in productive conversation can't duck this issue because so many beliefs have been tied to the endless quest for purpose and meaning. There is just no way around it.

I suppose I could prattle on about what I think other people should strive for in their lives. But I'm pretty sure I'm not qualified because my own life is still a work in progress. Perhaps when I achieve personal perfection and figure out all the secrets of the universe I'll feel confident enough to write a script for someone else's life. At the present time, however, I don't see a guy in the mirror who has the meaning of life all sorted out. What I do see is someone who found a way that is working pretty well for him so far. So I'm only comfortable saying that this may be what everyone needs to do. Find what works best for you and ride off into the sunset. That's the best cliché I can offer. Wherever your search leads, however, I am certain of one thing: Make sure to bring your brain along with you.

Please don't think that you are too weak to stand on your own without a bunch of unproven claims and beliefs propping you up and showing you the way. If that is the state you are in currently, then please rethink your course in life. Chances are you arrived at this conclusion only because of exposure to someone else's ideas about human strength and weakness. From birth, the not-so-subtle influences of your culture probably blasted you relentlessly with messages that numerous irrational beliefs were perfectly rational. Be careful about trusting the direction of your life on the words of fallible human beings. They may mean well when they invent and promote all these claims and beliefs, but don't forget that every one of them is playing the game with an imperfect and deceptive human brain. For this reason alone, they can't be trusted to be right about everything. Look at our history, the herd has been wrong many, many times before. The herd is wrong about many things right now, no doubt. Look around. Are you surrounded by infallible geniuses? Didn't think so. Honestly, what do you think the chances are that you just happened to have been born into the first family, first society, and first time period in all of history that got it right on everything important?

It's your life, and you have to decide how to live it at as you see fit, of course, but I hope you will at least take into consideration the abilities and potential of that magnificent thinking machine of yours. It can achieve wonders and protect and guide you well. But only if you decide to help it help you. With a well-cared-for brain that is set free, you can investigate, explore, imagine, and create your way through life without ever having to be a weak skeptic who passively embraces unproven beliefs. Good skeptics can still live on the edge if they so choose. Good skeptics can hope for the unlikely and reach for the impossible. Only they do it without pretending to know things they do not know. Their ignorance is a motivation to learn rather than a reason to lie.

Why would you want or need a belief to give your life meaning when it was likely only made up by someone anyway? Why not make your own meaning with your own brain? Why not choose your own path toward what *you* define as a meaningful life? Why not find confidence and comfort by looking within yourself and by reaching outside yourself to other human beings who know and share similar hopes and fears?

I will close this uncomfortable but necessary brief excursion into the meaning of life by encouraging you to take responsibility for yourself. The meaning of life, for you, is whatever *you* say it is. Don't trade years of your life for delusions or lies manufactured by other people who failed to think well or just want your money. Trust me, your brain will fool you more than enough all by itself. You don't need anyone else helping you to make mistakes.

PRESENT FOR THE SHOW

I hope that everyone, younger people especially, recognize the exciting times we live in. The rate of scientific progress is accelerating. Although the last century was one of unprecedented discovery and change, this current century seems certain to top it. Today the horizon feels closer, and it's more difficult than ever to glimpse what is coming. What we can be sure about is that the potential and the likelihood for profound change has never been more obvious. As a species, we have always improved, always been good at learning more and doing better. But now it's as if we have hit a higher gear and can't slow down. In recent decades, both the universe and our biology have come into sharp focus like never before. But so much more is sure to come. Computing power and related technology now seem close to breakthroughs that could cause profound changes to human culture overnight. I don't know if all this will fuel a spectacular voyage into a wonderful new future or burn down civilization forever. Either way, however, it's going to be interesting. From sending robots to distant worlds to mapping our own brain, we are doing more amazing things and doing them more frequently. Unfortunately, we also have to consider that increasingly dangerous and powerful weapons are likely to become available to more nations and more people in the coming decades. How will it turn out for us? This could be the early stages of a glorious golden age of discovery, a leap forward never to be matched again. This may be the very century that our descendants look back on thousands of years from now and identify as the moment when humankind finally woke up got busy. I understand that this might not feel like such a special time because of the wars, greed, poverty, violence, corruption, political nonsense, and other self-inflicted wounds that dominate headlines. But that stuff is not the whole story. We really are within sight of becoming an enlightened and positive species that we finally can be unreservedly proud of. You know, one that doesn't let nine million of its babies die in poverty each year while simultaneously spending trillions of dollars on weapons to destroy ourselves. We really can do better. We really can elevate humankind to a higher plane. It's reasonable to hope for. It's possible. But that doesn't mean it will happen tomorrow, or ever, of course.

Things could also go very badly for us. Given our deep loyalties to nations over humanity and belief over reason, we could just as easily tumble backward into a repeat of the Dark Ages. Sadly, some people are pushing hard in that direction right now. But there is no denying that there really is this *possibility* of an amazing and positive future for humankind. Given all the advancements seen in recent years—and the reasonable promise of computer power, robotics, genetics, and nano-technology—we do seem close to a game-changing moment. But what will it mean if our brains can't keep up with our toys? It will do us little good to build the modern world of our dreams if it is not inhabited by modern people. What achievement would it really be if we were to colonize the Moon and Mars, only to have half the people who live there spend their days hunting for Martian ghosts and worrying about how being on a different world might affect their horoscopes? What is the point of mapping the human genome and brain if the masses still prefer medical quackery? Progress must mean more than improved tools. It has to mean improved thinking, too.

Fortunately, the sweep of human history as a whole indicates that we just might be getting somewhere. Maybe. Science is now taught to and appreciated by more people than ever before, both in raw numbers and as a ratio to global population. The World Wide Web may be a significant problem in the way it facilitates the spread of irrational beliefs, but let's not forget that it has made science and reason more accessible to a greater number of people than ever before. Even in most societies with high rates of supernatural/paranormal belief today, skeptics and doubters are widely tolerated. There was a time not too long ago in our past when a person like me who speaks and writes about the need to question everything and challenge extraordinary claims would have been in great physical danger, not from random extremists, but from legitimate authority figures, backed up by the laws of the land and the majority of the public. Merely communicating some of the ideas in this book likely would have earned me a death sentence everywhere on Earth just a short time ago. Today, however, a copy of this book can be purchased and read by any citizen in most countries. We are moving in the right direction. We are becoming better thinkers and more appreciative of skepticism. Unfortunately, the pace is too slow. What may well happen is that tomorrow's global culture will be a muddled mess similar to what we have now, only more extreme. If we all don't start to become more skeptical and scientific in our thinking, we may end up with a world deeply spilt with science-minded visionaries on one side and stagnant magic-seekers on the other. It may be the doers versus the deniers, the scientific versus the superstitious. Don't wait too long to choose a side.

If we hope to ever become a true *thinking* species, sometime before the Sun flames out, then we must solve the crisis of weak skepticism. Collectively we have to think more and believe less. The amount of irrational thinking that passes for business as usual these days leaves humankind to face each new day blindfolded, with one arm tied behind its back, and one foot stuck in a bucket. This is no way to run a species. The world needs more good skeptics. Those who already are good skeptics have the ability—and, I suggest, a moral obligation—to spread the word and encourage more people to think well.

For readers who may be closet skeptics or new to the critical-thinking lifestyle, please don't ever allow yourself to become intimidated or discouraged by the swirling sea of madness around you. I know it can feel lonely, like those inevitable times when you are the only person in the room who doesn't think JFK was shot by aliens as predicted by a horoscope written by Nostradamus. We've all been there. I understand that it can feel overwhelming when it seems every thread of society is hopelessly soaked through with superstition. It's not strange for you to feel outnumbered, because you are. But maybe the numbers are changing in the favor of skeptics. We will have to wait and see. Regardless, you can make it as a good skeptic. You can succeed in society, have friends, raise a family, and do whatever you want to do with your life that makes the journey worthwhile. You never have to sacrifice your intellectual dignity. You can keep both feet planted firmly in reality. You have the tools. You have the power. Your human brain— properly maintained and applied—will keep you free and clear of most lies and most delusions most of the time.

It seems to me that a book has done well if it leaves readers with a few thoughts that remain potent and useful long after the book is closed and put away. At its best, a book should not only inform but motivate, inspire, and haunt as well. If any small cluster of words were to be fortunate enough to survive among your memories long after this book is closed, I hope it might be these:

Appreciate the magnificent brain you possess. Protect and nurture it. Strive to be a good skeptic so that few hours of your precious life will be squandered on dead-end beliefs. Always try to think like a scientist so that you might better know truth from fiction. When unusual claims and extraordinary beliefs come your way, challenge them. Question everything. Never flinch, never cave when faced with true mysteries. No matter how tempting, don't pretend to know things you do not know. Always, the right reaction is to think. Think before you leap. Think before you believe. Just keep thinking.

GOOD THINKING!

- Many people claim that various beliefs are necessary to find excitement, joy, and meaning in life. But the full and rewarding lives of many skeptics around the world prove that this is not necessarily true.
- Science is a never-ending process of exploration, discovery, and imagination. New information and ideas are always coming. New questions are always generated. Make sure your brain is ready to welcome it all.
- Learn and appreciate as much of the human story as possible. Accept it all—the good and the bad—as personal to you.

We are a young and closely related species. Therefore *all* history is *your* history.

- An amazing universe of microbial mysteries and gigantic galaxies offers more than enough to excite and inspire us. There simply is no reason to rely upon unproven claims and dubious beliefs to make our lives interesting.

RESOURCES TO KEEP LEARNING

CHAPTER 1: STANDING TALL ON A FANTASY-PRONE PLANET

Carroll, Robert Todd, ed. *The Skeptic's Dictionary*. Hoboken, NJ: John Wiley and Sons, 2003. Also available at www.skepdic.com.

Davis, Hank. *Caveman Logic: The Persistence of Primitive Thinking in a Modern World*. Amherst, NY: Prometheus Books, 2009.

Dunning, Brian. *Skeptoid: A Critical Analysis of Pop Phenomena*. Seattle, WA: Thunderwood Press, 2007.

———. *Skeptoid* (podcast), http://skeptoid.com.

James Randi Education Foundation, www.randi.org/site/.

Novella, Steven. *Neurologica* (blog), http://theness.com/neurologicablog/.

Sagan, Carl. *The Demon-Haunted World: Science as a Candle in the Dark*. New York: Random House, 1995.

Schick, Theodore, and Lewis Vaughn. *How to Think about Weird Things*. New York: McGraw-Hill, 2011.

Shermer, Michael. *The Believing Brain: From Ghosts and Gods to Politics and Conspiracies—How We Construct Beliefs and Reinforce Them as Truths*. New York: Times Books, 2011.

Skeptic (magazine), www.skeptic.com.

Skeptical Inquirer magazine, www.csicop.org/si.

Skepticality (official podcast of *Skeptic* magazine), www.skepticality.com.

Smith, Jonathan C. *Pseudoscience and Extraordinary Claims of the Paranormal: A Critical Thinker's Toolkit*. West Sussex, UK: Wiley-Blackwell, 2010.

Thompson, Damian. *Counterknowledge: How We Surrendered to Conspiracy Theories, Quack Medicine, Bogus Science, and Fake History*. New York: W. W. Norton, 2008.

CHAPTER 2: PAY A VISIT TO THE STRANGE THING THAT LIVES INSIDE YOUR HEAD

Brain Games. (DVD) National Geographic, 2011.

Brockman, John. *The Mind: Leading Scientists Explore the Brain, Memory, Personality, and Happiness*. New York: Harper Perennial, 2011.

———, ed. *This Will Make You Smarter: New Scientific Concepts to Improve Your Thinking*. New York: Harper Perennial, 2009.

Buonomano, Dean. *Brain Bugs: How the Brain's Flaws Shape Our Lives*. New York: W. W. Norton, 2011.

Chabris, Christopher, and Daniel Simons. *The Invisible Gorilla and Other Ways Our Intuitions Deceive Us*. New York: Crown, 2010.

DiSalvo, David. *What Makes Your Brain Happy and Why You Should Do the Opposite*. Amherst, NY: Prometheus Books, 2011.

McRaney, David. *You Are Not So Smart: Why You Have Too Many Friends on Facebook, Why Your Memory Is Mostly Fiction, and 46 Other Ways You're Deluding Yourself*. New York: Gotham, 2012. Also visit the companion website: http://youarenotsosmart.com/.

———. *You Are Now Less Dumb: How to Conquer Mob Mentality, How to Buy Happiness, and All the Other Ways to Outsmart Yourself*. New York: Gotham. 2013.

Ratey, John J. *A User's Guide to the Brain: Perception, Attention, and the Four Theaters of the Brain*. New York: Vintage, 2002.

Stanovich, Keith E. *How to Think Straight about Psychology*. New York: HarperCollins, 1996.

CHAPTER 3: A THINKER'S GUIDE TO UNUSUAL CLAIMS AND WEIRD BELIEFS

Clegg, Brian. *Armageddon Science*. New York: St. Martin's Press, 2010.

Committee for Skeptical Inquiry, www.csicop.org.

Debunker's Domain, www.debunker.com.

Feder, Kenneth L. *Encyclopedia of Dubious Archaeology: From Atlantis to the Walam Olum*. Santa Barbara, CA: Greenwood, 2010.

———. *Frauds, Myths, and Mysteries: Science and Pseudoscience in Archaeology*. Boston: McGraw Hill, 2008.

Guterl, Fred. *The Fate of the Species: Why The Human Race May Cause Its Own Extinction and What We Can Do about It*. New York: Bloomsbury, 2012.

Harrison, Guy P. *50 Popular Beliefs That People Think Are True.* Amherst, NY: Prometheus Books, 2012.

———. *50 Reaons People Give for Believing in a God.* Amherst, NY: Prometheus Books, 2008.

———. *50 Simple Questions for Every Christian.* Amherst, NY: Prometheus Books, 2013.

———. *Race and Reality: What Everyone Should Know about Our Biological Diversity.* Amherst, NY: Prometheus Books, 2010.

Jordan, Michael. *Dictionary of Gods and Goddesses.* New York: Facts on File, 2004.

Kelly, Lynne. *The Skeptic's Guide to the Paranormal.* New York: Avalon, 2004.

Kurtz, Paul. *The New Skepticism: Inquiry and Reliable Knowledge.* Amherst, NY: Prometheus Books, 1992.

———, ed. *Science and Religion: Are They Compatible?* Amherst, NY: Prometheus Books, 2003.

———, ed. *Skeptical Odysseys.* Amherst, NY: Prometheus Books, 2001.

Newitz, Annalee. *Scatter, Adapt, and Remember: How Humans Will Survive a Mass Extinction.* New York: Doubleday, 2013.

Nickell, Joe. *Adventures in Paranormal Investigation.* Lexington: University Press of Kentucky, 2007.

———. *Tracking the Man-Beasts: Sasquatch, Vampires, Zombies, and More.* Amherst, NY: Prometheus Books, 2011.

Nonprophets (radio show), www.nonprophetsradio.com.

Plait, Philip. *Bad Astronomy.* New York: John Wiley and Sons, 2002.

———. *Death from the Skies!* New York: Penguin Books, 2008.

Radford, Benjamin. *Scientific Paranormal Investigation: How to Solve the Unexplained Mysteries.* Corrales, NM: Rhombus, 2010.

Randi, James. *An Encyclopedia of Claims, Frauds, and Hoaxes of the Occult and Supernatural.* New York: St. Martin's Griffin, 1995.

———. *Flim-Flam!* Amherst, NY: Prometheus Books, 1982.

———. *The Mask of Nostradamus.* Amherst, NY: Prometheus Books, 1993.

Schaffer, Robert. *Bad UFOs* (blog), http://badufos.blogspot.com.

———. *UFO Sightings: The Evidence.* Amherst, NY: Prometheus Books, 1998.

Shermer, Michael. "Learn to Be Psychic in Ten Easy Lessons!" *Skeptic* (blog), www.skeptic.com/downloads/10_Easy_Psychic_Lessons.pdf, 2009.

———. *Science Friction: Where the Known Meets the Unknown.* New York: Times Books, 2005.

Singh, Simon, and Edzard Ernst. *Trick or Treatment: The Undeniable Facts about Alternative Medicine.* New York: W. W. Norton, 2008.

Barrett, Stephen. *Quackwatch* (guide to medical quackery/fraud), http://
quackwatch.org/.

CHAPTER 4: THE PROPER CARE AND FEEDING
OF A THINKING MACHINE

Blech, Jorg. *Healing through Exercise: Scientifically Proven Ways to Prevent
and Overcome Illness and Lengthen Your Life*. Cambridge, MA: Merloyd
Lawrence Book by Da Capo Press, 2009.

Carter, Rita. *The Human Brain Book*. New York: DK Adult, 2009.

Horstman, Judith. *The Scientific American: Brave New Brain*. San Francisco,
CA: Jossey-Bass, 2010.

———. *The Scientific American: Day in the Life of Your Brain*. San Francisco,
CA: Jossey-Bass, 2009.

Medina, John. *Brain Rules: 12 Principles for Surviving and Thriving at Work,
Home and School*. Seattle, WA: Pear Press, 2008.

Ratey, John J. *Spark: The Revolutionary New Science of Exercise and the Brain*.
Boston: Little, Brown, 2013.

Reynolds, Gretchen. *The First 20 Minutes: Surprising Science Reveals How We
Can Exercise Better, Train Smarter, Live Longer*. New York: Hudson Street
Press, 2012.

Shenk, David. *The Genius in All of Us*. New York: Doubleday, 2010.

Sweeney, Michael S. *Brain: The Complete Mind: How It Develops, How It Works,
and How to Keep It Sharp*. Washington, DC: National Geographic, 2009.

———. *Brainworks: The Mind-bending Science of How You See, What You
Think, and Who You Are*. Washington, DC: National Geographic, 2011.

CHAPTER 5: SO LITTLE TO LOSE AND A UNIVERSE TO GAIN

BBC History (website), www.bbc.co.uk/history/0/.

Becoming Human (website), www.becominghuman.org.

Becoming Human: Unearthing Our Earliest Human Ancestors (DVD), PBS,
2010.

Ben-Barak, Den. *The Invisible Kingdom: From the Tips of Our Fingers to the
Tops of Our Trash, Inside the Curious World of Microbes*. New York: Basic
Books, 2009.

Bill Nye the Science Guy (website), www.billnye.com.

Brockman, John, ed. *This Will Change Everything: Ideas That Will Shape the Future*. New York: Harper Perennial, 2009.

Brockman, John. *Science at the Edge: Conversations with the Leading Scientific Thinkers of Today*. New York: Union Square Press, 2008.

———, ed. *What Are You Optimistic About? Today's Leading Thinkers Lighten Up*. New York: Harper Perennial, 2007.

Broll, Brandon. *Microcosmos: Discovering the World through Microscopic Images from 20 X to Over 22 Million X Magnification*. Ontario, Canada: Firefly Books, 2010.

Calder, Nigel. *Magic Universe: A Grand Tour of Modern Science*. New York: Oxford University Press, 2003.

Clegg, Brian. *The God Effect: Quantum Entanglement, Science's Strangest Phenomenon*. New York: St. Martin's Griffin, 2006.

Dawkins, Richard. *The Magic of Reality: How We Know What's Really True*. New York: Free Press, 2011.

Discover (magazine), http://discovermagazine.com.

Dunn, Rob. *The Wild Life of Our Bodies: Predators, Parasites, and Partners That Shape Who We Are Today*. New York: Harper, 2011.

Encyclopedia of Life (website), http://eol.org.

For All Mankind (DVD), Criterion, 1989.

From the Earth to the Moon (DVDs), HBO Films and Tom Hanks, 2005.

Gonik, Larry. *Cartoon History of the Universe*. New York: Doubleday, 1997.

Hanlon, Michael. *Eternity: Our Next Billion Years*. Hampshire, UK: Palgrave Macmillan, 2008.

The History Place (website), www.historyplace.com.

How It Works (magazine), online at www.howitworksdaily.com.

Ingraham, John L. *March of the Microbes: Sighting the Unseen*. Reprint, Cambridge, MA: Belknap Press, 2012.

Kaku, Michio. *Physics of the Future: How Science Will Shape Human Destiny and Our Daily Lives by the Year 2100*. New York: Doubleday, 2011.

Kramer, Stephen (author) and Dennis Kunkel (photographer). *Hidden Worlds: Looking through a Scientist's Microscope*. London: Sandpiper, 2003.

Krauss, Lawrence M. *A Universe from Nothing: Why There Is Something Rather Than Nothing*. New York: Atria, 2013.

Kurtz, Paul. *Affirmations: Joyful and Creative Exuberance*. Amherst, NY: Prometheus Books, 2004.

Kurzweil, Ray. *The Singularity Is Near: When Humans Transcend Biology*. New York: Penguin, 2006.

Loxton, Daniel. *Evolution: How We and All Living Things Came to Be*. Tonawanda, NY: Kids Can Press, 2010.

NASA Images (website), www.nasa.gov/multimedia/imagegallery/index.html.

New Scientist (magazine), online at www.newscientist.com.

Nouvian, Claire, *The Deep: The Extraordinary Creatures of the Abyss*. Chicago: University of Chicago Press, 2007.

NOVA Science Now (website), www.pbs.org/wgbh/nova/sciencenow.

Nova: Fabric of the Cosmos, NOVA, PBS, 2011.

Popular Science (magazine), online at www.popsci.com.

Red Orbit (science news), online at www.redorbit.com.

Rees, Martin. *Our Final Hour*. New York: Basic Books, 2003.

———, ed. *Universe: The Definitive Visual Guide*. Reprint, London: DK, 2008.

Sagan, Carl. *Billions & Billions: Thoughts on Life and Death at the Brink of the Millennium*. New York: Ballantine, 1998.

Sagan, Carl. *Cosmos*. New York: Ballantine Books, 1985.

———. *Pale Blue Dot: A Vision of the Human Future in Space*. New York: Random House, 1994.

Sass, Erik, and Steve Wiegand. *The Mental Floss History of the World: An Irreverent Romp through Civilization's Best Bits*. New York: William Morrow Paperbacks, 2009.

Sawyer, G. J., and Victor Deak. *The Last Human: A Guide to Twenty-Two Species of Extinct Humans*. New Haven, CT: Yale University Press, 2007.

Science Illustrated (magazine), online at www.scienceillustrated.com.

Scientific American (magazine), online at www.scientificamerican.com.

Shermer, Michael. *The Borderlands of Science: Where Sense Meets Nonsense*. New York: Oxford University Press, 2002.

Shostak, Seth. *Confessions of an Alien Hunter: A Scientist's Search for Extraterrestrial Intelligence*. Washington, DC: National Geographic, 2009.

Shubin, Neil. *The Universe Within: Discovering the Common History of Rocks, Planets, and People*. New York: Pantheon, 2013.

Space.com (website), www.space.com/.

Stevenson, Mark. *An Optimist's Tour of the Future*. New York: Avery, 2011.

Stringer, Chris. *Lone Survivors: How We Came to Be the Only Humans on Earth*. New York: Times Books, 2011.

Tattersall, Ian. *Extinct Humans*. New York: Basic Books, 2001.

———. *Masters of the Planet: The Search for Our Human Origins*. New York: Palgrave Macmillan, 2012.

Through the Wormhole with Morgan Freeman (DVD), Revelations Entertainment, 2010.

Toomey, David. *Weird Life: The Search for Life That Is Very, Very Different from Our Own*. New York: W. W. Norton, 2013.

Trefil, James. *Space Atlas: Mapping the Universe and Beyond*. Washington, DC: National Geographic, 2012.

Tyson, Neil DeGrasse. *Space Chronicles: Facing the Ultimate Frontier*. New York: W. W. Norton, 2012.

Webb, Stephen. *If the Universe Is Teeming with Aliens . . . Where Is Everybody?* New York: Copernicus Books, 2002.

Weinberg, Steven. *Facing Up: Science and Its Cultural Adversaries*. Cambridge, MA: Harvard University Press, 2003.

———. *Lake Views: This World and the Universe*. Cambridge, MA: Belknap Press of Harvard University Press, 2010.

What Does It Mean to Be Human? (Smithsonian-run website), http://human origins.si.edu.

When We Left Earth: The NASA Missions (DVDs), Discovery Channel, 2008.

Wilson, Edward O. *The Diversity of Life*. Cambridge, MA: Belknap Press of Harvard University Press, 2010.

———. *Naturalist*. Washington, DC: Island Press, 1994.

Zimmer, Carl. *Evolution: The Triumph of an Idea*. New York: Harper Perennial, 2006.

———. *Parasite Rex: Inside the Bizarre World of Nature's Most Dangerous Creatures*. New York: Free Press, 2001.

———. *A Planet of Viruses*. Chicago: University of Chicago Press, 2011.

NOTES

CHAPTER 1: STANDING TALL ON A FANTASY-PRONE PLANET

1. Hank Davis, *Caveman Logic: The Persistence of Primitive Thinking in a Modern World* (Amherst, NY: Prometheus Books, 2009).

2. BBC News, "Nigeria 'Child Witch Killer' Held," BBC News, December 4, 2008, http://news.bbc.co.uk/2/hi/africa/7764575.stm (accessed July 17, 2013).

3. Salman Ravi, "Village 'Witches' Beaten," BBC News, October 20, 2009, http://news.bbc.co.uk/2/hi/south_asia/8315980.stm (accessed July 17, 2013); BBC News, "Indian 'Witchcraft' Family Killed," BBC News, March 19, 2006, http://news.bbc.co.uk/2/hi/south_asia/4822750.stm (accessed July 17, 2013).

4. James Gallagher, "Paralysed Woman's Thoughts Control Robotic Arm," BBC News, December 16, 2012, http://www.bbc.co.uk/news/health-20731973 (accessed July 12, 2013).

CHAPTER 2: PAY A VISIT TO THE STRANGE THING
THAT LIVES INSIDE YOUR HEAD

1. Joannie Schrof Fischer, "What Is Memory Made Of?" *Mysteries of Science* (*US News and World Report*), 2002, p. 27.

2. Ingfei Chen, *Scientific American*, September 6, 2011, www.scientific american.com/article.cfm?id=911-memory-accuracy (accessed February 11, 2013).

3. Michael Shermer, "Patternicity," *Skeptic*, December 2008, www.michael shermer.com/2008/12/patternicity/ (accessed February 12, 2013).

4. Ibid.

CHAPTER 3: A THINKER'S GUIDE TO UNUSUAL CLAIMS
AND WEIRD BELIEFS

1. David W. Moore, "Three in Four Americans Believe in Paranormal,"

Gallup News Service, June 16, 2005, http://www.gallup.com/poll/16915/three
-four-americans-believe-paranormal.aspx (accessed February 4, 2013).

2. Linda Lyons, "Paranormal Beliefs Come (Super) Naturally to Some,"
November 1, 2005, http://www.gallup.com/poll/19558/Paranormal-Beliefs
-Come-SuperNaturally-Some.aspx (accessed January 30, 2013).

3. Antonio Regalado, "Poll: Mexicans Express Belief in Spirits, Not
Science," January 5, 2011, http://news.sciencemag.org/scienceinsider/2011/01/
poll-mexicans-express-belief-in.html?ref=hp (accessed January 30, 2013).

4. Leagle, "State v. Perry," July 15, 2003, http://www.leagle.com/xml
Result.aspx?page=1&xmldoc=20031290582SE2d708_11159.xml&docbase
=CSLWAR 2-1986-2006&SizeDisp=7 (accessed February 23, 2013).

5. Pride Chigwedere, George R. Seage III, Sofia Gruskin, Tun-Hou Lee,
and M. Essex, "Estimating the Lost Benefits of Antiretroviral Drug Use in
South Africa," *Journal of Acquired Immune Deficiency Syndrome* 49, no.
4 (December 1, 2008): 410, http://www.aids.harvard.edu/Lost_Benefits.pdf
(accessed January 4, 2013).

6. V. A. Luyckx, V. Steenkamp, and M. J. Stewart, "Acute Renal Failure
Associated with the Use of Traditional Folk Remedies in South Africa," *Ren
Fail*, January 27, 2005, http://www.ncbi.nlm.nih.gov/pubmed/15717633 (ac-
cessed January 30, 2013).

7. I. A. Malik, S. Gopalan, "Use of CAM Results in Delay in Seeking
Medical Advice for Breast Cancer," *European Journal of Epidemiology*, August
18, 2003, www.ncbi.nlm.nih.gov/sites/entrez?cmd=Retrieve&db=pubmed&dopt
=AbstractPlus&list_uids=12974558 (accessed March 10, 2013).

8. Michael Shermer, "The Reality Distortion Field," *Skeptic* 17, no. 4, 2012.

9. Charlie Jones, "Design Thinking," CBS, January 6, 2013 (accessed
January 31, 2013).

10. Steve Kroft, "Steve Jobs," CBS, October 20, 2011, www.cbsnews.com/
video/watch/?id=7385390n (accessed January 31, 2013).

11. Bruce Hood, *The Science of Superstition: How the Developing Brain
Creates Supernatural Beliefs* (New York: HarperCollins Paperback, 2010), p.
157.

12. Katelyn Catanzariti, "Homeopath, Wife Jailed over Baby's Death,"
Age, September 28, 2009, http://news.theage.com.au/breaking-news-national/
homeopath-wife-jailed-over-babys-death-20090928-g8w4.html (accessed January 2, 2013).

13. Yusuke Fukui and Akiko Okazaki, "Homeopathy under Scrutiny after
Lawsuit over Death of Infant," *Asahi Shimbun*, September 6, 2010, http://www
.asahi.com/english/TKY201008050254.html (accessed November 5, 2012).

14. International Alliance for Animal Therapy and Healing, www.iaath
.com/treatments/flower.shtml (accessed March 14, 2013).

15. Coco Ballantyne, "Strange but True: Drinking Too Much Water Can Kill,"
Scientific American, June 21, 2007, http://www.scientificamerican.com/article
.cfm?id=strange-but-true-drinking-too-much-water-can-kill (accessed January
30, 2013).

16. Harris Poll, "What People Do and Do Not Believe In," Harris Interactive, December 15, 2009, http://www.harrisinteractive.com/vault/Harris
_Poll_2009_12_15.pdf (accessed January 10, 2013).

17. Lyons, "Paranormal Beliefs Come (Super) Naturally to Some."

18. Pew Forum on Religion and Public Life, "Many Americans Mix
Multiple Faiths," Pew Forum, December 9, 2009, http://pewforum.org/Other
-Beliefs-and-Practices/Many-Americans-Mix-Multiple-Faiths.aspx (accessed
February 27, 2013).

19. Kaja Perina, "Alien Abductions: The Real Deal?" *Psychology Today*,
March 1, 2003, www.psychologytoday.com/articles/%5Byyyy%5D%5Bmm%5
D/%5Btitle-raw%5D (accessed January 29, 2013).

20. Susan A. Clancy, *Abducted: How People Come to Believe They Were
Kidnapped by Aliens* (Cambridge, MA: Harvard University Press, 2005), p. 35.

21. Tom Head, ed., *Conversations with Carl Sagan* (Jackson: Univ. Press
of Mississippi, 2006), p. 101.

22. Christopher D. Bader, Carson F. Mencken, and Joseph O. Baker,
Paranormal America (New York: New York University Press, 2010), p. 106.

23. Greg Long, *The Making of Bigfoot: The Inside Story* (Amherst, NY:
Prometheus Books, 2004), p. 336.

24. Ibid. pp. 443—51.

25. Timothy Egan, "Search for Bigfoot Outlives the Man Who Created
Him," *New York Times*, January 3, 2003, http://www.nytimes.com/2003/01/03/
us/search-for-bigfoot-outlives-the-man-who-created-him.html (accessed January 7, 2013).

26. Wil S. Hylton, "Craig Venter's Bugs Might Save the World," *New York
Times*, May 30, 2012, http://www.nytimes.com/2012/06/03/magazine/craig
-venters-bugs-might-save-the-world.html?pagewanted=all&_r=1& (accessed
January 29, 2013).

27. Edward O. Wilson, *Naturalist* (Washington, DC: Island Press, 1994),
p. 364.

28. Moore, "Three in Four Americans Believe in Paranormal."

29. "RAAF Captures Flying Saucer in Roswell Region," *Roswell Daily
Record*, July 8, 1947, p. 1.

30. *Roswell Daily Record*, July 9, 1947, p. 1.

31. B. D. Gildenberg, "A Roswell Requiem," *Skeptic* 10, no. 1 (2003): 60–63.

32. Ibid.

33. Joe Kittinger, interview with the author; quoted in Guy P. Harrison, "I Was the First Man in Space," *Caymanian Compass*, October 26, 2001.

34. Annie Jacobson, *Area 51: An Uncensored History of America's Top Military Base* (New York: Little, Brown), 2011, pp. 367–74.

35. "Good Heavens! An Astrologer Dictating the President's Schedule?" *Time*, May 16, 1988, http://www.time.com/time/magazine/article/0,9171,967389 -1,00.html (accessed January 31, 2013).

36. Harris Poll, "What People Do and Do Not Believe In."

37. Pat Robertson, "Reinhard Bonnke Tells of Nigerian Man Raised from the Dead," *700 Club*, www.cbn.com/700club/features/bonnke_raisedpastor.aspx (accessed January 31, 2013).

38. Erich von Däniken, *Chariots of the Gods?* (Berkeley, CA: Berkley Books, 1999), p. 108.

39. Ibid, p. 96.

40. Ibid, p. 65.

41. Ibid, p. 73.

42. David Robson, "A Brief History of the Brain," *New Scientist*, September 24, 2011, p. 45.

43. Penn State University, "How Were the Egyptian Pyramids Built?" *ScienceDaily*, March 29, 2008, http://www.sciencedaily.com/releases/2008/03/ 080328104302.htm (accessed February 5, 2013).

44. Big Picture Science, *Zombies Aren't Real: Guy Harrison*, November 12, 2012, http://radio.seti.org/blog/2012/11/big-picture-science-zombies-arent-real -guy-harrison/ (accessed February 27, 2013).

45. Big Picture Science, *Doomsday Live!* part 1, http://radio.seti.org/ blog/2012/11/big-picture-science-doomsday-live-part-1/ (accessed February 27, 2013).

46. Jennifer Viegas, "Human Extinction: How Could It Happen?" *Discovery News*, http://news.discovery.com/human/human-extinction-doomsday.html (accessed September 5, 2011).

47. Jon Hamilton, "Psst! The Human Brain Is Wired for Gossip," NPR, *Morning Edition*, May 20, 2011, http://www.npr.org/2011/05/20/136465083/ psst-the-human-brain-is-wired-for-gossip (accessed January 26, 2013).

48. Frank Newport, "Landing a Man on the Moon: The Public's View," July 20, 1999, Gallup News Service, http://www.gallup.com/poll/3712/landing-man -moon-publics-view.aspx (accessed January 31, 2013).

49. "*Apollo 11* Hoax: One in Four People Do Not Believe in Moon Landing," *Telegraph*, July 17, 2009, http://www.telegraph.co.uk/science/space/5851435/ Apollo-11-hoax-one-in-four-people-do-not-believe-in-moon-landing.html (accessed January 3, 2011).

50. Christopher Hitchens, "Hugo Boss: What I Learned about Hugo Chávez's Mental Health When I Visited Venezuela with Sean Penn," *Slate*, March 25, 2013, www.slate.com/articles/news_and_politics/fighting_words/ 2010/08/hugo_boss.html (accessed March 12, 2013).

51. Mary Lynne Dittmar, "Engaging the 18–25 Generation: Educational Outreach, Interactive Technologies, and Space," Dittmar Associates, 2006, http://www.dittmar-associates.com/Publications/Engaging%20the%20 18-25%20Generation%20Update~web.pdf (accessed March 22, 2013).

52. C-SPAN, "Apollo 16 and Space Exploration," January 11, 2013, http:// www.c-spanvideo.org/program/310334-1 (accessed February 27, 2013).

53. As quoted by Andrew Chaikin in Guy P. Harrison, *50 Popular Beliefs That People Think Are True* (Amherst, NY: Prometheus Books, 2011), p. 89.

54. Newport, "Landing a Man on the Moon."

55. *Wikipedia*, s.v. "Hister," http://en.wikipedia.org/wiki/Hister#cite _note-1 (accessed August 5, 2013).

56. Guy P. Harrison, "God Is in This Place," *Caymanian Compass*, November 19, 1993, pp. 10–11.

57. Charles Berlitz, *The Bermuda Triangle* (New York: Avon, 1975).

58. Charles Berlitz, *Atlantis: The Lost Continent Revealed* (London: Fontana/Collins, 1985); and Charles Berlitz and William L. Moore, *The Roswell Incident* (New York: Berkley Books, 1980).

59. Larry Kusche, *The Bermuda Triangle Mystery—Solved* (Amherst, NY: Prometheus Books, 1995).

60. Kusche, *Bermuda Triangle Mystery—Solved*, pp. 275–77.

61. Naval Heritage and History Command, "The Bermuda Triangle," http://www.history.navy.mil/faqs/faq8-1.htm (accessed January 26, 2013).

62. "Does the Bermuda Triangle Really Exist?" United State Coast Guard, http://www.uscg.mil/history/faqs/triangle.asp (accessed January 25, 2013).

63. Baylor Institute for Studies of Religion, "American Piety in the 21st Century," Baylor University, September 2006, p. 45, http:// www.baylor.edu/ content/services/document.php/33304.pdf (accessed January 2, 2013).

64. Theodore Schick and Lewis Vaughn, *How to Think about Weird Things* (New York: McGraw-Hill, 2011), p. 7.

65. "Tsunami Clue to 'Atlantis' Found," BBC News, August 15, 2005, http:// news.bbc.co.uk/2/hi/science/nature/4153008.stm (accessed February 20, 2013).

66. Paul Rincon, "Satellite Images 'Show Atlantis,'" BBC News, June 6, 2004, http://news.bbc.co.uk/2/hi/science/nature/3766863.stm (accessed February 20, 2013).

67. "Atlantis 'Obviously Near Gibraltar,'" BBC News, September 20, 2001, http://news.bbc.co.uk/2/hi/science/nature/1554594.stm (accessed February 11, 2013).

68. Kenneth L. Feder, *Encyclopedia of Dubious Archaeology: From Atlantis to the Walam Olum* (Santa Barbara, CA: Greenwood, 2010), p. 33.

69. William B. Scott, "The Truth Is Out There: A Veteran Reporter Describes His Search for the Aircraft of Area 51," *Air & Space Magazine*, September 1, 2010, http://www.airspacemag.com/military-aviation/The-Truth -is-Out-There.html?c=y&page=1 (accessed January 23, 2013).

CHAPTER 4: THE PROPER CARE AND FEEDING OF A THINKING MACHINE

1. M. C. Morris, D. A. Evans, C. C. Tangney, J. L. Bienias, and R. S. Wilson, "Associations of Vegetable and Fruit Consumption with Age-Related Cognitive Change," *Neurology* 67, no. 8 (October 24, 2006): 1370–76, www.neurology. org/content/67/8/1370.abstract?sid=431a2aff-e7cd-441e-9ba3-94ba4575c92e (accessed February 1, 2013).

2. "Eating More Berries May Reduce Cognitive Decline in the Elderly," *ScienceDaily*, www.sciencedaily.com/releases/2012/04/120426110250.htm (accessed February 18, 2013).

3. For an excellent roundup of the science of exercise and how it can help your brain, read Gretchen Reynolds' *The First 20 Minutes: Surprising Science Reveals How We Can Exercise Better, Train Smarter, Live Longer* (New York: Hudson Street Press, 2012).

4. John Ratey, *A User's Guide to the Brain: Perception, Attention, and the Four Theaters of the Brain* (New York: Vintage, 2002), p. 359.

5. David Hinkley, "Americans Spend 34 Hours a Week Watching TV, According to Nielsen Numbers," *New York Daily News*, September 19, 2012, www.nydailynews.com/entertainment/tv-movies/americans-spend-34-hours -week-watching-tv-nielsen-numbers-article-1.1162285 (accessed February 14, 2013).

6. Centers for Disease Control and Prevention, "The Association between School-Based Physical Activity, Including Physical Education, and Academic Performance," July 2010, www.cdc.gov/healthyyouth/health_and_academics/ pdf/pa-pe_paper.pdf (accessed February 4, 2013), p. 6.

7. Alexandra Sifferlin, "Why Prolonged Sitting Is Bad for Your Health," *Time*, March 28, 2012, http://healthland.time.com/2012/03/28/standing-up-on-the-job-one-way-to-improve-your-health/ (accessed January 26, 2013).

8. John Medina, *Brain Rules* (Seattle, WA: Pear Press, 2008), p. 26.

9. Ibid, p. 5.

10. Reynolds, *First 20 Minutes*, p. 205.

11. "Facts and Figures," Drowsy Driving, http://drowsydriving.org/about/facts-and-stats/ (accessed February 11, 2013).

12. Katie Moisse, "5 Health Hazards Linked to Lack of Sleep," June 11, 2012, http://abcnews.go.com/Health/Sleep/health-hazards-linked-lack-sleep/story?id=16524313 (accessed February 2, 2013).

13. Medina, *Brain Rules*, pp. 152–53.

14. "Juggling Languages Can Build Better Brains," *ScienceDaily*, www.sciencedaily.com/releases/2011/02/110218092529.htm (accessed February 17, 2013).

15. "First Physical Evidence Bilingualism Delays Onset of Alzheimer's Symptoms," *ScienceDaily*, www.sciencedaily.com/releases/2011/10/111013121701.htm (accessed February 17, 2013).

16. "Juggling Enhances Connections in the Brain," *ScienceDaily*, www.sciencedaily.com/releases/2009/10/091016114055.htm (accessed February 17, 2013).

17. "Well-Connected Brains Make You Smarter in Older Age," *ScienceDaily*, www.sciencedaily.com/releases/2012/05/120523102958.htm (accessed January 3, 2013).

18. "Reading, Writing and Playing Games May Help Aging Brains Stay Healthy," *ScienceDaily*, www.sciencedaily.com/releases/2012/11/121125103947.htm (accessed February 17, 2013).

19. Corrie Goldman, "This Is Your Brain on Jane Austen, and Stanford Researchers Are Taking Notes," Stanford Reader, http://news.stanford.edu/news/2012/september/austen-reading-fmri-090712.html (accessed January 29, 2013).

CHAPTER 5: SO LITTLE TO LOSE AND A UNIVERSE TO GAIN

1. Linda Lyons, "Paranormal Beliefs Come (Super) Naturally to Some," Gallup News Service, November 1, 2005, http://www.gallup.com/poll/19558/Paranormal-Beliefs-Come-SuperNaturally-Some.aspx (accessed February 21, 2013).

2. David W. Moore, "Three in Four Americans Believe in Paranormal," Gallup News Service, June 16, 2005, http://www.gallup.com/poll/16915/three-four-americans-believe-paranormal.aspx (accessed December 27, 2012).

3. "The Global Religious Landscape," Pew Forum, December 18, 2012, http://www.pewforum.org/2012/12/18/global-religious-landscape-exec/ (accessed July 31, 2013).

4. Phil Zuckerman, "Atheism: Contemporary Numbers and Patterns," pp. 47–65 in *The Cambridge Companion to Atheism*, edited by Michael Martin (New York: Cambridge University Press, 2007).

5. "An Intellectual Entente," *Harvard Magazine*, September 10, 2009, http://harvardmagazine.com/breaking-news/james-watson-edward-o-wilson-intellectual-entente (accessed February 21, 2013).

6. Edward O. Wilson, *Naturalist* (Washington, DC: Island Press, 1994), p. 364.

7. "Loneliest Bug on Earth . . . Has a Friend," *New Scientist*, December 15, 2012, p. 20.

8. Rose Eveleth, "Up with Microbes," *Scientific American*, March 2013, p. 14.

9. Carl Zimmer, *A Planet of Viruses* (Chicago: University of Chicago Press, 2011), p. 42.

10. Debora MacKenzie, "Rosacea May Be Caused by Mite Faeces in Your Pores," *New Scientist*, August 30, 2012, www.newscientist.com/article/dn22227-rosacea-may-be-caused-by-mite-faeces-in-your-pores.html (accessed March 24, 2013).

11. Peter Nova, "The Star in You," *NOVA ScienceNOW*, December 2, 2010, http://www.pbs.org/wgbh/nova/space/star-in-you.html (accessed July 31, 2013).

BIBLIOGRAPHY

Bader, Christopher, F. Carson Mencken, and Joseph Baker. *Paranormal America*. New York: New York University Press, 2010.

Barker, Dan. *Maybe Yes, Maybe No: A Guide for Young Skeptics*. Amherst, NY: Prometheus Books, 1990.

Barrett, Stephen, and William T. Jarvis, eds. *The Health Robbers: A Close Look at Quackery in America*. Amherst, NY: Prometheus Books, 1993.

Bausell, R. Barker. *Snake Oil Science: The Truth about Complementary and Alternative Medicine*. Oxford: Oxford University Press, 2009.

Ben-Barak, Den. *The Invisible Kingdom: From the Tips of Our Fingers to the Tops of Our Trash, Inside the Curious World of Microbes*. New York: Basic Books, 2009.

Bennett, Jeffrey. *Beyond UFOs: The Search for Extraterrestrial Life and Its Astonishing Implications for Our Future*. Princeton, NJ: Princeton University Press, 2008.

Berlitz, Charles. *Atlantis: The Lost Continent Revealed*. London: Fontana/Collins, 1985.

Berlitz, Charles, and William L. Moore. *The Roswell Incident*. New York: Berkley Books, 1980.

Blackmore, Susan. *Beyond the Body*. Chicago: Academy Chicago Publishers, 1992.

———. *Dying to Live: Near-Death Experiences*. Amherst, NY: Prometheus Books, 1993.

———. *In Search of the Light: The Adventures of a Parapsychologist*. Amherst, NY: Prometheus Books, 1996.

Blech, Jorg. *Healing through Exercise: Scientifically Proven Ways to Prevent and Overcome Illness and Lengthen Your Life*. Cambridge, MA: Merloyd Lawrence Book by Da Capo Press, 2009.

Bostrom, Nick, and Milan M. Cirkvoic, eds. *Global Catastrophic Risks*. New York: Oxford University Press, 2008.

Brockman, John, ed. *Intelligent Thought: Science versus the Intelligent Design Movement*. New York: Vintage Books, 2006.

———. *The Mind: Leading Scientists Explore the Brain, Memory, Personality, and Happiness*. New York: Harper Perennial, 2011.

———. ed. *This Will Make You Smarter: New Scientific Concepts to Improve Your Thinking.* New York: Harper Perennial, 2009.

———. ed. *What Have You Changed Your Mind About?* New York: Harper Perennial, 2009.

Brockman, Max, ed. *What's Next? Dispatches from the Future of Science.* New York: Vintage, 2009.

Broll, Brandon. *Microcosmos: Discovering the World through Microscopic Images from 20 X to Over 22 Million X Magnification.* Ontario, Canada: Firefly Books, 2010.

Buonomano, Dean. *Brain Bugs: How the Brain's Flaws Shape Our Lives.* New York: W. W. Norton, 2011.

Calder, Nigel. *Magic Universe: A Grand Tour of Modern Science.* New York: Oxford University Press, 2003.

Carroll, Robert Todd, ed. *The Skeptic's Dictionary.* Hoboken, NJ: John Wiley and Sons, 2003.

Carter, Rita. *The Human Brain Book.* New York: DK Adult, 2009.

Chabris, Christopher, and Daniel Simons. *The Invisible Gorilla and Other Ways Our Intuitions Deceive Us.* New York: Crown, 2010.

Chaffe, John. *Thinking Critically.* Boston: Houghton Mifflin, 2000.

Clancy, Susan A. *Abducted: How People Come to Believe They Were Kidnapped by Aliens.* Cambridge, MA: Harvard University Press, 2005.

Clegg, Brian. *Armageddon Science.* New York: St. Martin's Press, 2010.

———. *The God Effect: Quantum Entanglement, Science's Strangest Phenomenon.* New York: St. Martin's Griffin, 2006.

Collins, Michael. *Carrying the Fire: An Astronaut's Journeys.* New York: Farrar, Straus and Giroux, 2009.

Darling, David. *Life Everywhere: The Maverick Science of Astrobiology.* New York: Basic Books, 2001.

Davies, Paul. *The Eerie Silence: Renewing Our Search for Alien Intelligence.* Boston: Houghton Mifflin Harcourt, 2010.

Davis, Hank. *Caveman Logic: The Persistence of Primitive Thinking in a Modern World.* Amherst, NY: Prometheus Books, 2009.

Davis, James C. *The Human Story: Our History, from the Stone Age to Today.* New York: Harper Perennial, 2005.

Dawkins, Richard. *The Ancestor's Tale.* Boston: Houghton Mifflin, 2004.

———. *The Blind Watchmaker: Why the Evidence of Evolution Reveals a Universe without Design.* New York: W. W. Norton, 1996.

———. *Climbing Mount Improbable.* New York: W. W. Norton, 1997.

———. *The Greatest Show on Earth: The Evidence for Evolution.* New York: Free Press, 2009.

———. *The Magic of Reality: How We Know What's Really True*. New York: Free Press, 2011.

DiSalvo, David. *What Makes Your Brain Happy and Why You Should Do the Opposite*. Amherst, NY: Prometheus Books, 2011.

Dunning, Brian. *Skeptoid: A Critical Analysis of Pop Phenomena*. Seattle, WA: Thunderwood Press, 2007.

———. *Skeptoid 2: More Critical Analysis of Pop Phenomena*. Seattle, WA: Skeptoid Media, 2008.

Ellis, Richard. *Imagining Atlantis*. New York: Vintage, 1999.

Epstein, Greg. *Good without God: What a Billion Nonreligious People Do Believe*. New York: Harper Paperbacks, 2010.

Feder, Kenneth L. *Encyclopedia of Dubious Archaeology: From Atlantis to the Walam Olum*. Santa Barbara, CA: Greenwood. 2010.

———. *Frauds, Myths, and Mysteries: Science and Pseudoscience in Archaeology*. Boston: McGraw-Hill, 2008.

Fernyhough, Charles. *Pieces of Light: How the New Science of Memory Illuminates the Stories We Tell About Our Pasts*. New York: Harper, 2013.

Ferris, Timothy. *The Whole Shebang: A State-of-the-Universe(s) Report*. New York: Simon and Schuster, 1998.

Fine, Cordelia. *A Mind of Its Own: How Your Brain Distorts and Deceives*. New York: W. W. Norton, 2006.

Forrest, Barbara, and Paul R. Gross. *Creationism's Trojan Horse: The Wedge of Intelligent Design*. Oxford: Oxford University Press, 2004.

Frazier, Kendrick. *Science under Siege: Defending Science, Exposing Pseudoscience*. Amherst, NY: Prometheus Books, 2009.

Freedman, Carl. *Conversations with Isaac Asimov*. Jackson: University Press of Mississippi, 2005.

Gardner, Martin. *Did Adam and Eve Have Navels: Debunking Pseudoscience*. New York: W. W. Norton, 2001.

———. *The New Age: Notes of a Fringe Watcher*. Amherst, NY: Prometheus Books, 1991.

———. *Science: Good, Bad, and Bogus*. Amherst, NY: Prometheus Books, 1989.

Goldacre, Ben. *Bad Science: Quacks, Hacks, and Big Pharma Flacks*. New York: Faber and Faber, 2010.

Gonik, Larry. *Cartoon History of the Universe*. New York: Doubleday, 1997.

Gottshcall, Jonathan. *The Storytelling Animal: How Stories Make Us Human*. New York: Houghton Mifflin Harcourt, 2012.

Greene, Brian. *The Elegant Universe: Superstrings, Hidden Dimensions, and the Quest for the Ultimate Theory*. New York: Vintage Books, 2000.

Gross, Mathew Barrett, and Mel Gilles. *The Last Myth*. Amherst, NY: Prometheus Books, 2012.

Guterl, Fred. *The Fate of the Species: Why the Human Race May Cause Its Own Extinction and What We Can Do about It*. New York: Bloomsbury, 2012.

Guyatt, Nicholas. *Have a Nice Doomsday: Why Millions of Americans Are Looking Forward to the End of the World*. United Kingdom: Ebury Press, 2007.

Halpern, Paul. *Countdown to Apocalypse: A Scientific Exploration of the End of the World*. Cambridge, MA: Perseus Publishing, 1998.

Hanlon, Michael. *Eternity: Our Next One Billion Years*. London: Macmillan, 2009.

Harrison, Guy P. *50 Popular Beliefs That People Think Are True*. Amherst, NY: Prometheus Books, 2012.

———. *50 Simple Questions for Every Christian*. Amherst, NY: Prometheus Books, 2013.

———. *50 Reasons People Give for Believing in a God*. Amherst, NY: Prometheus Books, 2008.

———. *Race and Reality: What Everyone Should Know about Our Biological Diversity*. Amherst, NY: Prometheus Books, 2010.

Haught, James A. *Holy Hatred*. Amherst, NY: Prometheus Books, 1995.

———. *Honest Doubt*. Amherst, NY: Prometheus Books, 2007.

———. *2,000 Years of Disbelief*. Amherst, NY: Prometheus Books, 1996.

Hazen, Robert M. *Genesis: The Scientific Quest for Life's Origins*. Washington, DC: Joseph Henry Press, 2007.

Head, Tom, ed. *Conversations with Carl Sagan*. Jackson: University Press of Mississippi, 2006.

Hedges, Chris. *Empire of Illusion: The End of Literacy and the Triumph of Spectacle*. New York: Nation Books, 2010.

Hemenway, Priya. *Hindu Gods*. San Francisco, CA: Chronicle Books, 2003.

Hines, Terrence. *Pseudoscience and the Paranormal*. Amherst, NY: Prometheus Books, 2003.

Horstman, Judith. *The Scientific American: Brave New Brain*. San Francisco, CA: Jossey-Bass, 2010.

———. *The Scientific American: Day in the Life of Your Brain*. San Francisco, CA: Jossey-Bass, 2009.

Humes, Edward. *Monkey Girl: Education, Education, Religion, and the Battle for America's Soul*. New York: HarperCollins, 2007.

Ingraham, John L. *March of the Microbes: Sighting the Unseen*. Reprint, Cambridge, MA: Belknap Press, 2012.

Jacobson, Annie. *Area 51: An Uncensored History of America's Top Military Base*. New York: Little, Brown, 2011.

Jayawardhana, Ray. *Strange New Worlds: The Search for Alien Planets and Life beyond Our Solar System*. Princeton, NJ: Princeton University Press, 2011.

Jordan, Michael. *Encyclopedia of Gods*. London: Kyle Cathie, 2002.

Kaku, Michio. *Physics of the Future: How Science Will Shape Human Destiny and Our Daily Lives by the Year 2100*. New York: Doubleday, 2011.

Kaminer, Wendy. *Sleeping with Extra-terrestrials*. New York: Pantheon Books, 1999.

Kaufman, Marc. *First Contact: Scientific Breakthroughs in the Hunt for Life beyond Earth*. New York: Simon and Schuster, 2011.

Kelly, Lynne. *The Skeptic's Guide to the Paranormal*. New York: Avalon, 2004.

Kida, Thomas. *Don't Believe Everything You Think*. Amherst, NY: Prometheus Books, 2006.

Klass, Philip J. *The Real Roswell Crashed-Saucer Coverup*. Amherst, NY: Prometheus Books, 1997.

Kramer, Stephen (author), and Dennis Kunkel (photographer). *Hidden Worlds: Looking through a Scientist's Microscope*. London: Sandpiper, 2003.

Krauss, Lawrence M. *A Universe from Nothing: Why There Is Something Rather Than Nothing*. New York: Atria, 2013.

Kurtz, Paul. *Affirmations: Joyful and Creative Exuberance*. Amherst, NY: Prometheus Books, 2004.

———. *The New Skepticism: Inquiry and Reliable Knowledge*. Amherst, NY: Prometheus Books, 1992.

———. ed. *Science and Religion: Are They Compatible?* Amherst, NY: Prometheus Books, 2003.

———. ed. *Skeptical Odysseys*. Amherst, NY: Prometheus Books, 2001.

———. *The Transcendental Temptation: A Critique of Religion and the Paranormal*. Amherst, NY: Prometheus Books, 1991.

Kurzweil, Ray. *The Singularity Is Near*. New York: Penguin, 2006.

Kusche, Larry. *The Bermuda Triangle Mystery—Solved*. Amherst, NY: Prometheus Books, 1995.

Leeming, David. *A Dictionary of Creation Myths*. New York: Oxford University Press, 1994.

Levy, Joel. *A Bee in a Cathedral and 99 Other Scientific Analogies*. Buffalo, NY: Firefly Books, 2011.

Long, Greg. *The Making of Bigfoot: The Inside Story*. Amherst, NY: Prometheus Books, 2004.

Loxton, Daniel. *Evolution: How We and All Living Things Came to Be.* Tonawanda, NY: Kids Can Press, 2010.

Macknic, Stephen L., and Susana Martinez-Conde. *Sleights of Mind.* New York: Henry Holt, 2010.

Margulis, Lynn, and Dorian Sagan. *Microcosmos: Four Billion Years of Microbial Evolution.* Berkeley: University of California Press, 1997.

Marks, Jonathan. *Why I Am Not a Scientist.* Berkley: University of California Press, 2009.

Mayr, Ernst. *What Evolution Is.* London: Weidenfeld and Nicolson, 2002.

McAndrew, James. *Roswell Report: Case Closed.* Grand Prairie, TX: Books Express Publishing, 2011.

McGuire, Bill. *A Guide to the End of the World.* Oxford: Oxford University Press, 2002.

McRaney, David. *You Are Now Less Dumb: How to Conquer Mob Mentality, How to Buy Happiness, and All the Other Ways to Outsmart Yourself.* New York: Gotham. 2013.

McRaney, David. *You Are Not So Smart: Why You Have Too Many Friends on Facebook, Why Your Memory Is Mostly Fiction, and 46 Other Ways You're Deluding Yourself.* New York: Gotham, 2012.

Medina, John. *Brain Rules: 12 Principles for Surviving and Thriving at Work, Home, and School.* Seattle, WA: Pear Press, 2008.

Mills, David. *Atheist Universe.* Berkeley, CA: Ulysses Press, 2006.

Mnookin, Seth. *The Panic Virus: A True Story of Medicine, Science, and Fear.* New York: Simon and Schuster, 2011.

Mooney, Chris, and Sheril Kirshenbaum. *Unscientific America: How Scientific Illiteracy Threatens Our Future.* New York: Basic Books, 2009.

Murdoch, Stephen. *IQ: A Smart History of a Failed Idea.* Hoboken, NJ: Wiley and Sons, 2007.

National Academy of Sciences. *Science, Evolution, and Creationism.* Washington, DC: National Academies Press, 2008.

Nelson, Kevin. *The Spiritual Doorway in the Brain.* New York: Dutton, 2011.

Newitz, Annalee. *Scatter, Adapt, and Remember: How Humans Will Survive a Mass Extinction.* New York: Doubleday, 2013.

Nickell, Joe. *Adventures in Paranormal Investigation.* Lexington: University Press of Kentucky, 2007.

———. *Looking for a Miracle: Weeping Icons, Relics, Stigmata, Visions & Healing Cures.* Amherst, NY: Prometheus Books, 1999.

———. *The Mystery Chronicles: More Real-Life X-Files.* Lexington: University Press of Kentucky, 2004.

———. *Psychic Sleuths: ESP and Sensational Cases*. Amherst, NY: Prometheus Books, 1994.

———. *Relics of the Christ*. Lexington: University Press of Kentucky, 2007.

———. *Tracking the Man-Beasts: Sasquatch, Vampires, Zombies, and More*. Amherst, NY: Prometheus Books, 2011.

Nisbett, Richard E. *Intelligence and How to Get It*. New York: W. W. Norton, 2009.

Nouvian, Claire, *The Deep: The Extraordinary Creatures of the Abyss*. Chicago: University of Chicago Press, 2007.

Offit, Paul A. *Autism's False Prophets: Bad Science, Risky Medicine, and the Search for a Cure*. New York: Columbia University Press, 2010.

———. *Deadly Choices: How the Anti-Vaccine Movement Threatens Us All*. New York: Basic Books, 2010.

Offit, Paul A., and Charlotte A. Moser. *Vaccines and Your Child: Separating Fact from Fiction*. New York: Columbia University Press, 2011.

Palmer, Douglas. *Origins: Human Evolution Revealed*. New York: Mitchell Beazley, 2010.

Park, Robert. *Voodoo Science: The Road from Fraud to Foolishness*. New York: Oxford University Press, 2000.

Pigliucci, Massimo. *Nonsense on Stilts: How to Tell Science from Bunk*. Chicago: University of Chicago Press, 2010.

Piper, Don. *90 Minutes in Heaven*. Grand Rapids, MI: Revell, 2004.

Plait, Philip. *Bad Astronomy*. New York: John Wiley and Sons, 2002.

———. *Death from the Skies!* New York: Penguin Books, 2008.

Prothero, Stephen. *Religious Literacy: What Every American Needs to Know— And Doesn't*. New York: HarperOne, 2007.

Radford, Benjamin. *Media Mythmakers: How Journalists, Activists and Advertisers Mislead Us*. Amherst, NY: Prometheus Books, 2003.

———. *Scientific Paranormal Investigation: How to Solve the Unexplained Mysteries*. Corrales, NM: Rhombus, 2010.

Randi, James. *An Encyclopedia of Claims, Frauds, and Hoaxes of the Occult and Supernatural*. New York: St. Martin's Griffin, 1995.

———. *The Faith Healers*. Amherst, NY: Prometheus Books, 1989.

———. *Flim-Flam!* Amherst, NY: Prometheus Books, 1982.

———. *The Mask of Nostradamus*. Amherst, NY: Prometheus Books, 1993.

Ratey, John J. *Spark: The Revolutionary New Science of Exercise and the Brain*. Little, Brown, 2013.

———. *A User's Guide to the Brain: Perception, Attention, and the Four Theaters of the Brain*. Vintage, 2002.

Rees, Martin. *Our Final Hour*. New York: Basic Books, 2003.

Reese, Martin, ed. *Universe: The Definitive Visual Guide*. Reprint, London: DK, 2008.

Reynolds, Gretchen. *The First 20 Minutes: Surprising Science Reveals How We Can Exercise Better, Train Smarter, Live Longer*. New York: Hudson Street Press, 2012.

Ryan, Craig. *Pre-Astronauts: Manned Ballooning on the Threshold of Space*. Annapolis, MD: US Naval Institute Press, 2003.

Sagan, Carl. *Billions & Billions: Thoughts on Life and Death at the Brink of the Millennium*. New York: Ballantine, 1998.

———. *Cosmos*. New York: Random House, 1980.

———. *The Demon-Haunted World: Science as a Candle in the Dark*. New York: Random House, 1995.

———. *Pale Blue Dot: A Vision of the Human Future in Space*. New York: Random House, 1994.

———. *The Varieties of Scientific Experience: A Personal View of the Search for God*. New York: Penguin, 2007.

Saler, Benson, Charles A. Ziegler, and Charles B. Moore. *UFO Crash at Roswell: The Genesis of a Modern Myth*. Washington, DC: Smithsonian Books, 2010.

Sass, Erik, and Steve Wiegand. *The Mental Floss History of the World: An Irreverent Romp through Civilization's Best Bits*. New York: William Morrow Paperbacks, 2009.

Sawyer, G. J., and Victor Deak. *The Last Human: A Guide to Twenty-Two Species of Extinct Humans*. New Haven, CT: Yale University Press, 2007.

Schick, Theodore, and Lewis Vaughn. *How to Think about Weird Things*. New York: McGraw-Hill, 2011.

Scott, Eugenie C. *Evolution vs. Creationism: An Introduction*. Berkeley: University of California Press, 2009.

Sharpiro, Rose. *Suckers: How Alternative Medicine Makes Fools of Us All*. London: Harvill Secker, 2008.

Sheaffer, Robert. *UFO Sightings: The Evidence*. Amherst, NY: Prometheus Books, 1998.

Shenk, David. *The Genius in All of Us*. New York: Doubleday, 2010.

Shermer, Michael. *The Believing Brain: From Ghosts and Gods to Politics and Conspiracies—How We Construct Beliefs and Reinforce Them as Truths*. New York: Times Books, 2011.

———. *The Borderlands of Science: Where Sense Meets Nonsense*. New York: Oxford University Press, 2002.

———. *Science Friction: Where the Known Meets the Unknown*. New York: Times Books, 2005.

———. *Why Darwin Matters: The Case against Intelligent Design*. New York: Times Books, 2006.

———. *Why People Believe Weird Things*. New York: MJF Books, 1997.

Shostak, Seth. *Confessions of an Alien Hunter: A Scientist's Search for Extraterrestrial Intelligence*. Washington, DC: National Geographic, 2009.

Shubin, Neil. *The Universe Within: Discovering the Common History of Rocks, Planets, and People*. New York: Pantheon, 2013.

Singh, Simon, and Edzard Ernst. *Trick or Treatment: The Undeniable Facts about Alternative Medicine*. New York: W. W. Norton, 2008.

Smith, Cameron M. *The Fact of Evolution*. Amherst, NY: Prometheus Books, 2011.

Smith, Cameron M., and Charles Sullivan. *The Top 10 Myths about Evolution*. Amherst, NY: Prometheus Books, 2006.

Smith, Jonathan C. *Pseudoscience and Extraordinary Claims of the Paranormal: A Critical Thinker's Toolkit*. West Sussex, UK: Wiley-Blackwell, 2010.

Soutwood, Richard. *The Story of Life*. New York: Oxford University Press, 2004.

Specter, Michael. *Denialism: How Irrational Thinking Hinders Scientific Progress, Harms the Planet, and Threatens Our Lives*. New York: Penguin Press, 2009.

Stanovich, Keith E. *How to Think Straight about Psychology*. New York: HarperCollins, 1996.

Stenger, Victor J. *The Fallacy of Fine-Tuning: Why the Universe Is Not Designed for Us*. Amherst, NY: Prometheus Books, 2011.

———. *The New Atheism: Taking a Stand for Science and Reason*. Amherst, NY: Prometheus Books, 2009.

Streiber, Whitley. *Communion*. New York: Beech Tree Books, 1987.

Stringer, Chris. *Lone Survivors: How We Came to Be the Only Humans on Earth*. New York: Times Books, 2011.

Sweeney, Michael S. *Brain: The Complete Mind: How It Develops, How It Works, and How to Keep It Sharp*. Washington, DC: National Geographic, 2009.

———. *Brain Works: The Mind-bending Science of How You See, What You Think, and Who You Are*. Washington, DC: National Geographic, 2011.

Tattersall, Ian. *Extinct Humans*. New York: Basic Books, 2001.

———. *Masters of the Planet: The Search for Our Human Origins*. New York: Palgrave Macmillan, 2012.

Thompson, Damian. *Counterknowledge: How We Surrendered to Conspiracy Theories, Quack Medicine, Bogus Science, and Fake History*. New York: W. W. Norton, 2008.

Toomey, David. *Weird Life: The Search for Life That Is Very, Very Different from Our Own*. New York: W. W. Norton, 2013.

Trefil, James. *Space Atlas: Mapping the Universe and Beyond*. Washington, DC: National Geographic, 2012.

Tyson, Neil DeGrasse. *Death by Black Hole: And Other Cosmic Quandaries*. New York: W. W. Norton, 2007.

———. *Space Chronicles: Facing the Ultimate Frontier*. New York: W. W. Norton, 2012.

Van Hecke, Madeleine. *Blind Spots: Why Smart People Do Dumb Things*. Amherst, NY: Prometheus Books, 1997.

Wanjek, Christopher. *Bad Medicine: Misconceptions and Misuses Revealed, from Distance Healing to Vitamin O*. New York: Wiley, 2002.

Ward, Peter D., and Donald Brownlee. *The Life and Death of Planet Earth*. New York: Henry Holt, 2004.

Webb, Stephen. *If the Universe Is Teeming with Aliens . . . Where Is Everybody?* New York: Copernicus Books, 2002.

Weinberg, Steven. *Facing Up: Science and Its Cultural Adversaries*. Cambridge, MA: Harvard University Press, 2003.

———. *Lake Views: This World and the Universe*. Cambridge, MA: Belknap Press of Harvard University Press, 2010.

Wilson, Edward O. *The Diversity of Life*. Cambridge, MA: Belknap Press of Harvard University Press, 2010.

———. *Naturalist*. Washington, DC: Island Press, 1994.

Wiseman, Richard. *Paranormality: Why We See What Isn't There*. London: Macmillan, 2011.

Woerlee, G. M. *Mortal Minds: The Biology of Near-Death Experiences*. Amherst, NY: Prometheus Books, 2005.

Young, Matt, and Taner Edis. *Why Intelligent Design Fails: A Scientific Critique of the New Creationism*. Piscataway, NJ: Rutgers University Press, 2006.

Zimmer, Carl. *Evolution: The Triumph of an Idea*. New York: Harper Perennial, 2006.

———. *A Planet of Viruses*, Chicago: University of Chicago Press, 2011.

Zuckerman, Phil. *Faith No More: Why People Reject Religion*. New York: Oxford University Press, 2011.

———. *Society without God: What the Least Religious Nations Can Tell Us about Contentment*. New York: NYU Press, 2010.

INDEX